Die „Ergebnisse der Mathematik" erscheinen in einzelnen Heften von 5 bis 7 Bogen Umfang. Je 5 Hefte bilden in der Reihenfolge ihres Erscheinens einen Band.

Die auf der Innenseite der Seitenüberschriften angebrachten *kursiven* Zahlen *[5]* sind die Seitenzahlen des Bandes.

Jedes Heft der „Ergebnisse" ist einzeln käuflich. Bei Verpflichtung zum Bezug eines vollständigen Bandes tritt eine 10%ige Preisermäßigung ein. Die Bezieher des „Zentralblatt für Mathematik" erhalten, sofern sie sich zum Bezug eines ganzen Bandes verpflichten, auf den ermäßigten Bandpreis einen weiteren Nachlaß von 20%.

Verlagsbuchhandlung Julius Springer.

ERGEBNISSE DER MATHEMATIK
UND IHRER GRENZGEBIETE
HERAUSGEGEBEN VON DER SCHRIFTLEITUNG
DES
„ZENTRALBLATT FÜR MATHEMATIK"
ZWEITER BAND
5

THE THEORY OF MATRICES

BY

C. C. MAC DUFFEE

BERLIN
VERLAG VON JULIUS SPRINGER
1933

ISBN 978-3-642-98421-1 ISBN 978-3-642-99234-6 (eBook)
DOI 10.1007/978-3-642-99234-6

ERGEBNISSE DER MATHEMATIK
UND IHRER GRENZGEBIETE

HERAUSGEGEBEN VON DER SCHRIFTLEITUNG
DES
„ZENTRALBLATT FÜR MATHEMATIK“

ZWEITER BAND

MIT 4 FIGUREN

BERLIN
VERLAG VON JULIUS SPRINGER
1933

Inhaltsverzeichnis.

Preface.

Matric algebra is a mathematical abstraction underlying many seemingly diverse theories. Thus bilinear and quadratic forms, linear associative algebra (hypercomplex systems), linear homogeneous transformations and linear vector functions are various manifestations of matric algebra. Other branches of mathematics as number theory, differential and integral equations, continued fractions, projective geometry etc. make use of certain portions of this subject. Indeed, many of the fundamental properties of matrices were first discovered in the notation of a particular application, and not until much later recognized in their generality.

It was not possible within the scope of this book to give a completely detailed account of matric theory, nor is it intended to make it an authoritative history of the subject. It has been the desire of the writer to point out the various directions in which the theory leads so that the reader may in a general way see its extent. While some attempt has been made to unify certain parts of the theory, in general the material has been taken as it was found in the literature, the topics discussed in detail being those in which extensive research has taken place.

For most of the important theorems a brief and elegant proof has sooner or later been found. It is hoped that most of these have been incorporated in the text, and that the reader will derive as much pleasure from reading them as did the writer.

Acknowledgment is due Dr. LAURENS EARLE BUSH for a critical reading of the manuscript.

CYRUS COLTON MACDUFFEE.

Contents.

I. Matrices, Arrays and Determinants.

1. Linear algebra. A linear algebra $\mathfrak{A}$ of order n over a field $\mathfrak{F}$ is composed of two or more numbers $\alpha, \beta, \gamma, \ldots$ and three operations, addition $(+)$, multiplication $(\cdot)$ and scalar multiplication such that $\alpha + \beta$, $\alpha \cdot \beta$, αa, $a \alpha$ are uniquely defined numbers of $\mathfrak{A}$, where a is in $\mathfrak{F}$. It is further assumed that addition is commutative and associative, and that multiplication is distributive with respect to addition. If a and b are in $\mathfrak{F}$ it is assumed that

$$a\alpha = \alpha a\,, \quad a(b\alpha) = (ab)\,\alpha\,, \quad (a\alpha)\,(b\beta) = (ab)\,(\alpha\beta)\,,$$
$$(a+b)\,\alpha = a\alpha + b\alpha\,, \quad a(\alpha+\beta) = a\alpha + a\beta\,.$$

Finally it is assumed that $\mathfrak{A}$ contains a finite number of numbers $\varepsilon_1, \varepsilon_2, \ldots, \varepsilon_n$ such that every number of $\mathfrak{A}$ is of the form

$$a_1 \varepsilon_1 + a_2 \varepsilon_2 + \cdots + a_n \varepsilon_n\,,$$

where the a's are in $\mathfrak{F}$[1].

2. Representation by ordered sets. If

$$\alpha = a_1 \varepsilon_1 + a_2 \varepsilon_2 + \cdots + a_n \varepsilon_n$$

is a number of $\mathfrak{A}$, and if the ordered set

$$[a_1, a_2, \ldots, a_n]$$

of numbers of $\mathfrak{F}$ be made to correspond to α, addition, multiplication and scalar multiplication for such sets can be so defined that they give a representation of $\mathfrak{A}$. Thus if

$$\alpha \backsim [a_1, a_2, \ldots, a_n]\,, \qquad \beta \backsim [b_1, b_2, \ldots, b_n]\,,$$

then by definition

$$\alpha + \beta \backsim [a_1 + b_1,\, a_2 + b_2,\, \ldots,\, a_n + b_n]\,,\; a\alpha \backsim [a a_1,\, a a_2,\, \ldots,\, a a_n]\,.$$

If

$$\varepsilon_i \cdot \varepsilon_j = c_{ij1}\,\varepsilon_1 + c_{ij2}\,\varepsilon_2 + \cdots + c_{ijn}\,\varepsilon_n\,,$$

then

$$\alpha \cdot \beta = \sum a_i \varepsilon_i \cdot \sum b_j \varepsilon_j = \sum (a_i b_j) \cdot (\varepsilon_i \varepsilon_j) = \sum a_i\, b_j\, c_{ijk}\, \varepsilon_k$$

so that

$$\alpha \cdot \beta \backsim \left[\sum a_i\, b_j\, c_{ij1}\,,\; \sum a_i\, b_j\, c_{ij2}\,,\; \ldots,\; \sum a_i\, b_j\, c_{ijn}\right].$$

For every choice of the n^3 numbers c_{ijk} of $\mathfrak{F}$ an algebra over $\mathfrak{F}$ is obtained[2].

[1] DICKSON, L. E.: Algebren und ihre Zahlentheorie, p. 23. Zürich 1927.
[2] HAMILTON, W. R.: Elements of Quaternions.

If in particular the algebra is associative, then

$$(\varepsilon_i \cdot \varepsilon_j) \cdot \varepsilon_k = \varepsilon_i \cdot (\varepsilon_j \cdot \varepsilon_k). \qquad (i, j, k = 1, 2, \ldots, n)$$

This means that the n^3 numbers c_{ijk} are subject to the n^4 conditions

$$(2.1) \qquad \Sigma_k c_{ikr} c_{jsk} = \Sigma_k c_{ijk} c_{ksr}. \qquad (i, j, r, s = 1, 2, \ldots, n)$$

If R_i denotes the ordered set of numbers

$$R_i = (c_{isr}) = \left\| \begin{matrix} c_{i11} & c_{i21} & \cdots & c_{in1} \\ c_{i12} & c_{i22} & \cdots & c_{in2} \\ \cdot & \cdot & \cdot & \cdot \\ c_{i1n} & c_{i2n} & \cdots & c_{inn} \end{matrix} \right\|,$$

and if multiplication and scalar multiplication of such sets are defined by

$$R_i R_j = (c_{isr})(c_{jsr}) = (\Sigma_k c_{ikr} c_{jsk}),$$
$$a R_i = (a c_{isr}),$$

then (2.1) may be written[1]

$$R_i R_j = \Sigma_k c_{ijk} R_k.$$

Thus the sets R_i combine under multiplication in the same manner as the basis numbers ε_i of $\mathfrak{A}$. If the sets R_i are not linearly independent with respect to $\mathfrak{F}$, the sets R_i obtained by bordering R_i above with a row of 0's and on the left with δ_{ri} (KRONECKER's δ), are linearly independent and are isomorphic with the ε_i under multiplication[2].

If, finally, addition of sets is defined by the identity

$$R_i + R_j = (c_{isr} + c_{jsr}),$$

it is evident that these sets give a representation of the algebra[3].

It is interesting that in the same year that POINCARÉ's note was published, SYLVESTER wrote: "The PEIRCES (subsequently to 1858) had prefigured the universalization of HAMILTON's theory, and had emitted an opinion to the effect that probably all systems of algebraical symbols subject to the associative law of multiplication would be eventually found to be identical with linear transformations of schemata susceptible of matricular representation. ... That such must be the case it would be rash to assert, but it is very difficult to conceive how the contrary can be true[4]."

3. Total matric algebra. The considerations of § 2 suggest that a linear algebra of order n^2 over $\mathfrak{F}$ can be so defined that every algebra of order n over $\mathfrak{F}$ will be isomorphic with one of its proper sub-algebras[5].

[1] SCHUR, I.: Über eine Klasse von Matrizen, p. 59. Berlin 1901.

[2] DICKSON: Algebren und ihre Zahlentheorie, p. 35.

[3] POINCARÉ, H.: C. R. Acad. Sci., Paris Vol. 99 (1884) pp. 740—742. — WEYR, E.: S.-B. böhm. Ges. Wiss. Prague (1887) pp. 616—618. — STUDY, E.: Enc. math. Wiss. I A Vol. 4 (1904) § 10.

[4] SYLVESTER: Amer. J. Math. Vol. 6 (1884) pp. 270—286.

[5] VAN DER WAERDEN: Moderne Algebra Vol. I p. 37. Berlin 1930.

More generally we define the total matric algebra $\mathfrak{M}$ of order n^2 over a ring $\mathfrak{R}$ to consist of the sets of n^2 number each of the type

$$A = \left\| \begin{matrix} a_{11} & a_{12} & \dots & a_{1n} \\ a_{21} & a_{22} & \dots & a_{2n} \\ \dots & \dots & \dots & \dots \\ a_{n1} & a_{n2} & \dots & a_{nn} \end{matrix} \right\| = (a_{rs})$$

subject to the following operations and postulates:

Two sets $A = (a_{rs})$ and $B = (b_{rs})$ are equal if and only if $a_{rs} = b_{rs}$ for every r and s.

The operation of addition is defined by

$$A + B = (a_{rs} + b_{rs}) .$$

Evidently the sets form an abelian group with respect to addition, since the same is true of the elements of the ring $\mathfrak{R}$. The identity set, whose elements are all 0, will be denoted by O.

The operation of multiplication is defined by

$$AB = (\Sigma_i\, a_{ri}\, b_{is}) ,$$

that is, by "row by column" multiplication of the sets. The product is evidently unique, and $OA = AO = O$ for every A.

Multiplication in $\mathfrak{M}$ is associative, since multiplication in $\mathfrak{R}$ is associative.

$$\begin{aligned}(AB)\,C &= (\Sigma_{i,j}(a_{ri}\, b_{ij})\, c_{js}) \\ &= (\Sigma_{ij}\, a_{ri}(b_{ij}\, c_{js})) = A\,(BC) .\end{aligned}$$

Multiplication in $\mathfrak{M}$ is distributive with respect to addition, for the same is true in $\mathfrak{R}$.

$$\begin{aligned}(A + B)\,C &= (\Sigma_i(a_{ri} + b_{ri})\, c_{is}) \\ &(\Sigma_i(a_{ri}\, c_{is} + b_{ri}\, c_{is})) - AC + BC .\end{aligned}$$

Similarly $C\,(A + B) = CA + CB$.

Theorem 3.1. The total matric algebra of order n^2 over a ring $\mathfrak{R}$ is itself a ring.

Nothing is gained in this connection by specializing $\mathfrak{R}$, for multiplication in $\mathfrak{M}$ remains usually non-commutative even when $\mathfrak{R}$ is a commutative ring, and the inverse as to multiplication of $A \neq O$ does not always exist even for a ring without divisors of zero. If, however, $\mathfrak{R}$ is a ring with the unit element 1, then the matrix $I = (\delta_{rs})$ (KRONECKER's delta) is such that $AI = IA = A$ for every A, and $\mathfrak{M}$ is a ring with unit element I.

We define a *matrix* over $\mathfrak{R}$ to be a number of a total matric algebra over $\mathfrak{R}$.

Suppose $A = (a_{rs})$, $B = (b_{rs})$, $AB = C = (c_{rs})$ to be n-rowed matrices. Let A and B be separated into blocks:

$$A = \left\| \begin{matrix} A_{11} & A_{12} \\ A_{21} & A_{22} \end{matrix} \right\| , \qquad B = \left\| \begin{matrix} B_{11} & B_{12} \\ B_{21} & B_{22} \end{matrix} \right\| , \qquad C = \left\| \begin{matrix} C_{11} & C_{12} \\ C_{21} & C_{22} \end{matrix} \right\| ,$$

where A_{11} has i rows and j columns, B_{11} has j rows and k columns, and C_{11} has i rows and k columns. It is readily verified that

$$C_{rs} = \Sigma_i A_{ri} B_{is}, \qquad (r, s, i = 1, 2)$$

where the rectangular blocks A_{ij}, B_{ij} are multiplied "row by column". This holds for all separations of the matrices into blocks, provided the rows of B are separated in the same ways as the columns of A.

We shall use the word *array* to mean an ordered set of k elements such that two arrays are equal if and only if each consists of the same number of elements and if corresponding elements are equal. If $k = mn$, the array can be arranged in the form of a rectangle. Under certain circumstances the sum or product of two arrays may have meaning, for instance when they are matrices or as in the last paragraph when they are blocks cut from matrices, but no such operations are implicit in the definition of array.

A matrix is an instance of an array of n^2 elements, but it is much more than that, for it is a member of a total matric algebra for which the operations of addition and multiplication are defined. The importance of the matric theory derives from the rules of combination of matrices, while the fact that they may be represented by square arrays is incidental.

Sir Thomas Muir recently remarked: "One of the first of such extensions of usage was entirely uncalled for, especially in England, namely to make it (the word *matrix*) take the place already satisfactorily occupied by the word *array*. How satisfactory this was will be readily seen on looking through textbooks of determinants like Scott's[1]. A capable historian ... would certainly add further to his credit if in the course of his work he made manifest by precept and example an irreproachable mode of using in each other's company the terms *array*, *determinant*, *matrix*[2]."

Writers are far from agreed on a consistent terminology. The word "matrix" was first used by Sylvester[3] to denote a rectangular array from which determinants can be formed. The concept of a matrix as a hypercomplex number is due in essence to Hamilton but more directly to Cayley[4]. Châtelet, among others, uses "matrix" for a rectangular array and "tableau" for a member of a matric algebra[5]. But the essential point, to which Châtelet agrees, is to differentiate the concepts.

---

[1] Scott, R. F.: A treatise on the theory of determinants. Cambridge 1880.

[2] Muir, Thomas: Trans. Roy. Soc. S. Africa Vol. 18 III (1929) pp. 219–227.

[3] Sylvester: Philos. Mag. Vol. 37 (1850) pp. 363–370 — Coll. Works Vol. I p. 145.

[4] Cayley: Trans. London Phil. Soc. Vol. 148 (1858) pp. 17–37 — Coll. Works Vol. II pp. 475–496.

[5] Châtelet: Les groupes abéliens finis. Paris 1924.

4. Diagonal and scalar matrices. A matrix of the type

$$D = \left\| \begin{matrix} k_1 & 0 & 0 & \dots & 0 \\ 0 & k_2 & 0 & \dots & 0 \\ 0 & 0 & k_3 & \dots & 0 \\ \cdot & \cdot & \cdot & \cdot & \cdot \\ 0 & 0 & 0 & \dots & k_n \end{matrix} \right\| = [k_1, k_2, \dots, k_n]$$

is called *diagonal*. From the definitions of addition and multiplication of matrices it follows that

$$[k_1, k_2, \dots, k_n] + [l_1, l_2, \dots, l_n] = [k_1 + l_1, k_2 + l_2, \cdots, k_n + l_n],$$

$$[k_1, k_2, \dots, k_n][l_1, l_2, \dots, l_n] = [k_1 l_1, k_2 l_2, \dots, k_n l_n].$$

A diagonal matrix all of whose diagonal elements are equal is called *scalar*[1].

If S_k denotes $[k, k, \dots, k]$, then $S_k + S_l = S_{k+l}$, $S_k S_l = S_{kl}$, $S_0 = O$.

Thus

Theorem 4.1. The scalar matrices of $\mathfrak{M}$ constitute a subring of $\mathfrak{M}$ isomorphic with $\mathfrak{R}$.

It is customary to write k for S_k and kA for $S_k A$. If $\mathfrak{R}$ is a ring with unit element 1, then $S_1 = I$.

Theorem 4.2. If $\mathfrak{R}$ is a commutative ring with unit element, and if $AX = XA$ for every X in $\mathfrak{M}$, then A is scalar.

For if x_{hk} in

$$\sum a_{ri} x_{is} = \sum x_{ri} a_{is}$$

is replaced by $\delta_{hp} \delta_{kq}$, there results

$$\sum a_{ri} \delta_{ip} \delta_{sq} = \sum \delta_{rp} \delta_{iq} a_{is}, \quad a_{rp} \delta_{sq} = \delta_{rp} a_{qs}. \quad (r, s, p, q = 1, 2, \dots, n)$$

For $q = s$ and $r \neq p$ this gives $a_{rp} = 0$, while for $q = s$ and $r = p$ it gives $a_{rr} = a_{ss}$.

5. Transpose. Symmetric and skew matrices. The matrix $A^{\mathrm{T}} = (a_{sr})$ obtained from $A = (a_{rs})$ by changing rows to columns is called the *transpose*[2] of A. A matrix S such that $S^{\mathrm{T}} = S$ is called *symmetric*, a matrix Q such that $Q^{\mathrm{T}} = -Q$ is called *skew*[3] (*skew-symmetric*, or *alternating*).

---

[1] SYLVESTER: Amer. J. Math. Vol. 6 (1884) pp. 270–286 — Coll. Works Vol. IV pp. 208–224.

[2] Or conjugate. Many different notations for the transpose have been used, as A', $\bar{A}$, $\breve{A}$, A^*, A_1, ${}_tA$. The present notation is in keeping with a systematic notation which, it is hoped, may find favor.

[3] CAYLEY: J. reine angew. Math. Vol. 32 (1846) pp. 119–123 — Coll. Works Vol. I pp. 332–336. LAGUERRE: J. École polytechn. Vol. 25 (1867) pp. 215 to 264 — Œuvres Vol. I pp. 228–233.

Theorem 5.1.

$$(A + B + \cdots + K)^{\mathrm{T}} = A^{\mathrm{T}} + B^{\mathrm{T}} + \cdots + K^{\mathrm{T}}.$$

Theorem 5.2. If $\mathfrak{R}$ is a commutative ring $(AB \ldots K)^{\mathrm{T}} = K^{\mathrm{T}} \ldots B^{\mathrm{T}} A^{\mathrm{T}}$.

For $(AB)^{\mathrm{T}} = (\sum a_{ri} b_{is})^{\mathrm{T}} = (\sum b_{ir} a_{si}) = B^{\mathrm{T}} A^{\mathrm{T}}$. The general theorem follows by induction[1].

Theorem 5.3. If $\mathfrak{R}$ is a ring in which $2x = a$ is solvable for every a, every matrix of $\mathfrak{M}$ over $\mathfrak{R}$ is uniquely expressible as a sum of a symmetric and a skew matrix[2].

Assume
$$A = S + Q, \quad S^{\mathrm{T}} = S, \quad Q^{\mathrm{T}} = -Q.$$

Then $A^{\mathrm{T}} = S^{\mathrm{T}} + Q^{\mathrm{T}} = S - Q$ so that $S = \dfrac{A + A^{\mathrm{T}}}{2}$, $Q = \dfrac{A - A^{\mathrm{T}}}{2}$. Conversely for every A it is possible to form a symmetric matrix S and a skew matrix Q in the above manner.

6. Determinants. Let $\mathfrak{M}$ be an algebra of n-rowed matrices with elements in a field $\mathfrak{F}$. It is desirable to have associated with every matrix A of $\mathfrak{M}$ a number $\theta(A)$ of $\mathfrak{F}$ which would serve the purpose of an absolute value. This end would be attained by finding a scalar function $\theta(A)$ of the elements a_{rs} of a general matrix A such that,

1. For every A, $\theta(A)$ is a non-constant rational integral function of the a_{rs} of lowest degree such that

2. $\theta(AB) = \theta(A)\,\theta(B)$.

It follows directly from (2) by taking $B = I$ that $\theta(A) = \theta(A)\,\theta(I)$, and since $\theta(A)$ is not constant, $\theta(I) = 1$. If $B = O$, (2) gives $\theta(O) = \theta(A)\,\theta(O)$. Again because $\theta(A)$ is not constant, $\theta(O) = 0$.

When $n = 3$, for instance, let

$$V = \begin{Vmatrix} 0 & 1 & 0 \\ 1 & 0 & 0 \\ 0 & 0 & 1 \end{Vmatrix}, \qquad W(t) = \begin{Vmatrix} 1 & 0 & 0 \\ 0 & 1 & t \\ 0 & 0 & 1 \end{Vmatrix}, \qquad T(t) = \begin{Vmatrix} 1 & 0 & 0 \\ 0 & t & 0 \\ 0 & 0 & 1 \end{Vmatrix}.$$

$$V^2 = I, \qquad W(t)\,W(-t) = I, \qquad T(t)\,T(1/t) = I.$$

Since $\theta(W(t))\,\theta(W(-t)) = 1$, it follows that $\theta(W(t))$ is independent of t, otherwise the left member of the above equation would be of degree >0 in t. Hence $\theta(W(t))$ has for every value of t the same value that it has for $t = 0$, namely

$$\theta(W(t)) = 1.$$

Consider

$$\begin{Vmatrix} 0 & 1 & 0 \\ 1 & 0 & 0 \\ 0 & 0 & 1 \end{Vmatrix} \begin{Vmatrix} t & 0 & 0 \\ 0 & 1 & 0 \\ 0 & 0 & 1 \end{Vmatrix} \begin{Vmatrix} 0 & 1 & 0 \\ 1 & 0 & 0 \\ 0 & 0 & 1 \end{Vmatrix} = \begin{Vmatrix} 1 & 0 & 0 \\ 0 & t & 0 \\ 0 & 0 & 1 \end{Vmatrix}$$

---

[1] Cayley: Philos. Trans. Roy. Soc. London Vol. 148 (1858) pp. 17—37 — Coll. Works Vol. II pp. 475—496.

[2] Frobenius: J. reine angew. Math. Vol. 84 (1878) pp. 1—63. — Cayley: l. c.

or $VT_1V = T_2$. Since $\theta(V) = \pm 1$, $\theta(T_1) = \theta(T_2)$. Since $\theta(T(t))\,\theta(T(1/t)) = 1$ and $\theta(T(t))$ is of the same degree λ in t that $\theta(T(1/t))$ is in $1/t$, it follows that $\theta(T(t))$ must be a monomial $a_\lambda t^\lambda$ in t. Since $\theta(T(1)) = 1$, $a_\lambda = 1$. We now invoke the minimum principle (1) and assume that $\lambda = 1$. This is justified by actually obtaining under this restriction a function satisfying the given requirements. Then

$$\theta(T_i) = t\,.$$

Consider

$$\begin{Vmatrix} 1 & 0 & 0 \\ -1 & 1 & 0 \\ 0 & 0 & 1 \end{Vmatrix} \begin{Vmatrix} 1 & 1 & 0 \\ 0 & 1 & 0 \\ 0 & 0 & 1 \end{Vmatrix} \begin{Vmatrix} 1 & 0 & 0 \\ -1 & 1 & 0 \\ 0 & 0 & 1 \end{Vmatrix} \begin{Vmatrix} 0 & 1 & 0 \\ 1 & 0 & 0 \\ 0 & 0 & 1 \end{Vmatrix} = \begin{Vmatrix} 1 & 0 & 0 \\ 0 & -1 & 0 \\ 0 & 0 & 1 \end{Vmatrix},$$

or $W_1W_2W_1V = T(-1)$. Hence

$$\theta(V) = -1\,.$$

If A is a general matrix, $\theta(TA) = t\,\theta(A)$ so that

1′. The function $\theta(A)$ is a polynomial in the elements a_{ij} which is homogeneous and linear in the elements of each row.

Since $\theta(VA) = -\theta(A)$, it follows that

2′. $\theta(A)$ merely changes sign when two rows are permuted.

3′. $\theta(I) = 1$[1].

The three properties just stated were called by WEIERSTRASS the *characteristic properties* of a determinant. By 1′.

$$\theta(A) = \sum_{h_1, h_2, \ldots, h_n = 1}^{n} \varepsilon_{h_1 h_2 \ldots h_n} a_{1h_1} a_{2h_2} \ldots a_{nh_n}\,.$$

Permute rows 1 and 2 and add, whence by 2′,

$$\sum (\varepsilon_{h_1 h_2 \ldots h_n} + \varepsilon_{h_2 h_1 \ldots h_n})\, a_{1h_1} a_{2h_2} \ldots a_{nh_n} = 0\,.$$

Hence in general $\varepsilon_{h_1 h_2 \ldots h_n}$ is 0 if two subscripts coincide, while if the h's and k's are all distinct, $\varepsilon_{h_1 h_2 \ldots h_n} = \pm \varepsilon_{k_1 k_2 \ldots k_n}$ according as the substitution

$$\begin{pmatrix} h_1\, h_2 \ldots h_n \\ k_1\, k_2 \ldots k_n \end{pmatrix}$$

is even or odd. Since $\theta(I) = 1$, it follows that $\varepsilon_{1,2,\ldots,n} = 1$. Hence in general

(6.1) $$\theta(A) = \sum \varepsilon_{h_1 h_2 \ldots h_n} a_{1h_1} a_{2h_2} \ldots a_{nh_n}\,,$$

where the summation is over all permutations $(h_1, h_2, \ldots, h_n)$ of $(1, 2, \ldots, n)$, and $\varepsilon_{h_1 h_2 \ldots h_n}$ is 1 or -1 according as the permutation is even or odd[2].

---

[1] This treatment is due to K. HENSEL: J. reine angew. Math. Vol. 159 (1928) pp. 246—254.

[2] WEIERSTRASS-GÜNTHER: Zur Determinantentheorie. 1886—1887 — Werke Vol. III pp. 271—286. — KRONECKER: Vorlesungen über die Theorie der Determinanten Vol. 1 p. 291 et. seq. Teubner 1903.

That this function $\theta(A)$ satisfies the demands of HENSEL will follow from Corollary (7.9).

It is possible to develop the entire theory of determinants from the characteristic properties of WEIERSTRASS[1].

7. Properties of determinants. As CAYLEY remarked, "the idea of matrix (or square array) precedes that of determinant"[2]. A determinant is a number associated in a precise way with an array of n^2 ordered numbers. This point of view seems to have been clear to CAUCHY who gave the first systematic treatment of determinants, a treatment which can scarcely be improved upon today[3]. Unfortunately the word "determinant" is often used today to mean both an array and a number associated with that array. (Note the remarks of MUIR in § 3, and BENNETT's criticism of BÔCHER[4]. For a very clear statement of the ordinary determinant theorems from the present point of view, see HASSE, Höhere Algebra I, de Gruyter 1926.)

There is no room in the present tract for an extended treatment of determinant theory. A practically complete bibliography is given in MUIR's "Theory of determinants in the historical order of development", 4 v., Macmillan 1906—23. The early history is attractively presented by KRONECKER, "Vorlesungen über die Theorie der Determinanten", Teubner 1903, p. 1—9. An excellent reference book is G. KOWALEWSKI's "Einführung in die Determinantentheorie", Leipzig 1909.

Let A be a square array with elements in a field $\mathfrak{F}$ and let $d(A)$ or $|A|$ be its determinant. A few important theorems are listed for future reference.

Theorem 7.1. $d(A^{\mathrm{T}}) = d(A)$.

Theorem 7.2. If B is obtained from A by the interchange of two rows or of two columns, $d(B) = -d(A)$.

Theorem 7.3. If two rows or two columns of A are identical, $d(A) = 0$.

The usual proof of this result, namely in noting that the hypothesis gives $2d(A) = 0$ and hence $d(A) = 0$, fails when $\mathfrak{F}$ has the characteristic 2. The proof can be modified to include this case[5] or the result can be proved otherwise.

Theorem 7.4. If B is obtained from A by multiplying any row or any column of A by k, then $d(B) = kd(A)$.

Theorem 7.5. If A is a square array each element of whose kth row is a sum

$$d_{ks1} + d_{ks2} + \cdots + d_{ksm} \qquad (s = 1, 2, \ldots, n)$$

[1] ILIOVICI: Rev. Math. spéc. Vol. 37 (1927) pp. 433—436 and 457—458.
[2] CAYLEY: J. reine angew. Math. Vol. 50 (1855) pp. 282—285.
[3] CAUCHY: J. École polytechn. Vol. 10 (1815) pp. 51—112.
[4] BENNETT, A. A.: Amer. Math. Monthly Vol. 32 (1925) pp. 182—185.
[5] RYCHLIK, K.: J. reine angew. Math. Vol. 167 (1932) p. 197.

then

$$d(A) = d(A_1) + d(A_2) + \cdots + d(A_m),$$

where A_h is the array obtained by replacing the elements of the kth row by $d_{k1h}, d_{k2h}, \ldots, d_{knh}$. Similarly for columns.

Theorem 7.6. If B is obtained from A by adding to any row (or column) a linear combination of the other rows (columns), then $d(B) = d(A)$.

If A is a square array of n rows,

$$A^{r_1 \ldots r_k}_{s_1 \ldots s_k} = \begin{vmatrix} a_{r_1 s_1} & a_{r_1 s_2} & \cdots & a_{r_1 s_k} \\ a_{r_2 s_1} & a_{r_2 s_2} & \cdots & a_{r_2 s_k} \\ \cdot & \cdot & \cdot & \cdot \\ a_{r_k s_1} & a_{r_k s_2} & \cdots & a_{r_k s_k} \end{vmatrix}$$

is called an r-rowed *minor array* of A. $A^{r_1 \ldots r_k}_{r_1 \ldots r_k}$ *is a principal* minor array.

Theorem 7.7.

$$d(A) = \sum \pm d\left(A^{r_1 \ldots r_k}_{i_1 \ldots i_k}\right) d\left(A^{r_{k+1} \ldots r_n}_{i_{k+1} \ldots i_n}\right),$$

where the summation is over the $\binom{n}{k}$ ways of selecting the k numbers $i_1, \ldots, i_k$ from among the numbers $1, \ldots, n$ without regard to order, and the sign is + or − according as the substitution

$$\begin{pmatrix} 1, & 2, & \ldots, & k, & k+1, & \ldots, & n \\ r_1, & r_2, & \ldots, & r_k, & r_{k+1}, & \ldots, & r_n \end{pmatrix}$$

is even or odd[1].

In particular $d(A^r_s) = a_{rs}$. Let A_{rs} denote $\pm d\left(A^{1, \ldots, r-1, r+1, \ldots, n}_{1, \ldots, s-1, s+1, \ldots, n}\right)$, the sign being + or − according as

$$\begin{pmatrix} r, & 1, & \ldots, & r-1, & r+1, & \ldots, & n \\ s, & 1, & \ldots, & s-1, & s+1, & \ldots, & n \end{pmatrix}$$

is even or odd. Call A_{rs} the *cofactor* of a_{rs} in A.

Theorem 7.8.

$$\sum_{i=1}^{n} a_{pi} A_{qi} = \sum_{i=1}^{n} a_{ip} A_{iq} = \delta_{pq}\, d(A),$$

where δ_{pq} is KRONECKER's *delta.*

Theorem 7.9. Let A and B be n-rowed matrices with elements in a field $\mathfrak{F}$. Let $M^{k_1 \ldots k_m}_{l_1 \ldots l_m}$ be an m-rowed minor matrix of the product AB. Then

$$d\left(M^{k_1 \ldots k_m}_{l_1 \ldots l_m}\right) = \sum d\left(A^{k_1 \ldots k_m}_{i_1 \ldots i_m}\right) d\left(B^{i_1 \ldots i_m}_{l_1 \ldots l_m}\right),$$

where the summation is over all $\binom{n}{m}$ selections of $i_1, \ldots, i_m$ from $1, \ldots, n$ without regard to order[2].

Corollary 7.9. $d(AB) = d(A)\, d(B)$.

[1] LAPLACE: Mém. Acad. Sci. Paris 1772.

[2] DICKSON: Modern algebraic theories, p. 49. Chicago 1926.

8. Rank and nullity. If A is an $m \cdot n$ array with elements in a field $\mathfrak{F}$, the *rank* ϱ of A is the order of a minor square array of A of maximum order whose determinant is not zero[1]. If A is square of order n, then $n - \varrho$ is called the *nullity* of A.[2]

Theorem 8.1. If A is a square array of order n, with elements in a field $\mathfrak{F}$, of nullity $\varkappa$ and rank $\varrho = n - \varkappa$, there exist exactly $\varkappa$ linearly independent linear relations among the rows (columns) of A, and conversely.

Rearrange the columns of A so that the first r are linearly independent while each remaining column is linearly dependent upon these. Call the resulting matrix B. Then evidently $\varrho(A) = \varrho(B)$. There are exactly $n - r$ independent relations

$$k_{h1} b_{k1} + k_{h2} b_{k2} + \cdots + k_{h\varrho} b_{k\varrho} = b_{kh}$$
$$(k = 1, 2, \ldots, n;\ h = r + 1, \ldots, n)$$

among the columns of B. By Theorems 7.5 and 7.4 every $(r+1)$-rowed minor of A has a determinant which can be written as a linear combination of determinants each having at least two equal columns, and hence vanishes. Thus $\varrho \leq r$.

Now suppose that

$$\begin{vmatrix} b_{11} & \cdots & b_{1\varrho} \\ \cdot & \cdots & \cdot \\ b_{\varrho 1} & \cdots & b_{\varrho\varrho} \end{vmatrix} \neq 0 \qquad \begin{vmatrix} b_{11} & \cdots & b_{1\varrho} & b_{1h} \\ \cdot & \cdots & \cdot & \cdot \\ b_{\varrho 1} & \cdots & b_{\varrho\varrho} & b_{\varrho h} \\ b_{k1} & \cdots & b_{k\varrho} & b_{kh} \end{vmatrix} = 0$$

for every h and k. By Theorem 7.8

$$b_{p1} B_{k1}^{(h)} + b_{p2} B_{k2}^{(h)} + \cdots + b_{p\varrho} B_{k\varrho}^{(h)} + b_{ph} B_{kh}^{(h)} = 0$$

for every p and k. But the cofactors $B_{ki}^{(h)}$ are independent of k, so there is a relation

$$b_{p1} C_{1h} + b_{p2} C_{2h} + \cdots + b_{p\varrho} C_{\varrho h} + b_{ph} C_{hh} = 0$$

for every p and h. Moreover for $h > \varrho$, $C_{hh} \neq 0$ since B is of rank ϱ. Thus the number $n - r$ of linearly independent linear relations among the columns of B is at least $n - \varrho$, so $r \leq \varrho$. Hence $\varrho = r$, $n - r = \varkappa$.

Corollary 8.2. If $BC = O$, $\varrho(B) + \varrho(C) \leq n$. For every B there exists a C such that $\varrho(B) + \varrho(C) = n$.

A solution X of the equation $AX = O$ of rank $n - \varrho(A)$ is called a *complete solution*. Denote by (x) the *vector* or one-column array $(x_1, x_2, \ldots, x_n)$. Then $(y) = A(x)$ can be written for

$$\Sigma_j a_{ij} x_j = y_i . \qquad (i, j = 1, 2, \ldots, n) .$$

[1] FROBENIUS: J. reine angew. Math. Vol. 86 (1879) pp. 1–19. The concept of rank seems to be implicit, however, in a paper by I. HEGER: Denkschr. Akad. Wiss. Wien Vol. 14 (1858) pt. 2 pp. 1–121.

[2] SYLVESTER: Johns Hopkins Univ. Circulars Vol. III (1884) pp. 9–12 — Coll. Works Vol. IV pp. 133–145.

If B is a complete solution of the equation $AB = O$, and if (x) is an arbitrary vector, then $(y) = B(x)$ is the most general solution of the equation $A(y) = 0$.

Lemna 8.3. If the vector (z) ranges over the solutions of the equation $AB(z) = 0$, then $B(z)$ represents exactly $\varrho(B) - \varrho(AB)$ linearly independent vectors.

For among the $\tau = n - \varrho(AB)$ independent solutions of the equation $AB(z) = 0$ belong the $\sigma = n - \varrho(B)$ solutions $(z'), (z''), \ldots, (z^{(\sigma)})$ of the equation $B(z) = 0$. If these are extended by the solutions $(z^{(\sigma+1)}), \ldots, (z^{(\tau)})$ to a complete system of τ solutions, the $\tau - \sigma = \varrho(B) - \varrho(AB)$ vectors $B(z^{(\sigma+1)}), \ldots, B(z^{(\tau)})$ are independent. For a relation

$$C_{\sigma+1} B(z^{(\sigma+1)}) + \cdots + C_\tau B(z^{(\tau)}) = B(C_{\sigma+1}(z^{(\sigma+1)}) + \cdots + C_\tau(z^{(\tau)})) = O$$

would imply

$$C_{\sigma+1}(z^{(\sigma+1)}) + \cdots + C_\tau(z^{(\tau)}) = C_1(z') + \cdots + C_\sigma(z^{(\sigma)}),$$

and hence $C_1 = \cdots = C_\tau = O$.

Theorem 8.3. If A, B, C are three matrices of order n with elements in $\mathfrak{F}$,

$$\varrho(AB) + \varrho(BC) \leqq \varrho(B) + \varrho(ABC).$$

In the lemma replace B by BC. Then $\lambda = \varrho(BC) - \varrho(ABC)$ is the number of vectors $(y'), (y''), \ldots$ for which $ABC(y) = 0$ and the vectors $BC(y'), BC(y''), \ldots$ are independent. Then $(z') = C(y')$, $(z'') = C(y''), \ldots$ satisfy the equation $AB(z) = 0$, and the vectors $B(z') = BC(y'), B(z'') = BC(y''), \ldots$ are independent. But since there are not more than $\varrho(B) - \varrho(AB)$ such vectors[1] $(z'), (z''), \ldots,$

$$\lambda \leqq \varrho(B) - \varrho(AB).$$

Corollary 8.3. The nullity of the product of two matrices is at least as great as the nullity of either factor, and at most as great as the sum of the nullities of the factors[2].

For when $C = O$, FROBENIUS' theorem gives $\varrho(AB) \leqq \varrho(B)$; when $A = O$ it gives $\varrho(BC) \leqq \varrho(B)$; and when $B = I$, it gives

$$\varrho(A) + \varrho(C) \leqq n + \varrho(AC).$$

If A is an n-rowed square matrix with elements in a domain of integrity $\mathfrak{D}$ containing the field $\mathfrak{F}$ as a sub-variety, the column-nullity of A with respect to $\mathfrak{F}$ is the number of linearly independent linear relations among the columns of A with coefficients in $\mathfrak{F}$. The column-nullity may not equal the row-nullity[3].

[1] FROBENIUS: S.-B. preuß. Akad. Wiss. 1911 pp. 20—29.

[2] SYLVESTER's "Law of nullity". Johns Hopkins Univ. Circulars Vol. 3 (1884) pp. 9—12 — Coll. Works Vol. IV pp. 133—145.

[3] MACDUFFEE, C. C.: Ann. of Math. II Vol. 27 (1925) pp. 133—139.

Theorem 8.4. If A is a matrix of rank ϱ, and $d\left(A_{j_1\ldots j_\varrho}^{i_1\ldots i_\varrho}\right)$ is denoted by b_{ij} $\left(i, j = 1, 2, \ldots, m; m = \binom{n}{\varrho}\right)$, then the m-rowed square matrix $B = (b_{rs})$ is of rank 1.[1]

Suppose the rows and columns of A arranged so that $b_{11} \neq 0$. The last $n - \varrho$ rows are linear combinations of the first ϱ rows, so if suitable linear combinations of the first ϱ rows are added to each of the last $n - \varrho$ rows, these latter may be made to consist exclusively of 0's. This process does not alter the rank of A or the rank of B. Now $b_{rs} = 0$ except when $r = 1$, so the ratio

$$b_{i1} : b_{i2} : \cdots : b_{im}$$

is independent of i. In other words,

$$b_{rs} = k_r b_{is}, \qquad b_{11} \neq 0.$$

Theorem 8.5. In a symmetric matrix of rank ϱ there is at least one non-singular principal minor of order ϱ.[2]

Suppose A symmetric of rank ϱ. By Theorem 8.4 set

$$b_{ij} = m_i p_j = b_{ji} = m_j p_i, \qquad (i = 1, 2, \ldots, m)$$

where some p_i, say p_k, is not 0. Set $b_{kk}/p_k^2 = m$. Then $b_{ij} = m p_i p_j$. If m were 0, A would be of rank $<\varrho$. Hence $b_{kk} \neq 0$.

Theorem 8.6. The rank of a skew matrix is even[3].

Suppose A of odd rank ϱ and $b_{rs} = -b_{sr}$. Set

$$b_{ij} = m_i p_j, \quad b_{ji} = m_j p_i = -m_i p_j, \qquad p_k \neq 0.$$

For $m = -b_{kk}/p_k^2$ this implies

$$b_{ij} = m p_i p_j = -m p_j p_i.$$

Hence every $b_{ij} = 0$, contrary to the assumption that the rank of A was ϱ.[4]

9. Identities among minors. *Theorem 9.1. Among the minors of order m of a symmetric matrix A of order n there exists the relation*[5]

$$|a_{rs}| = \Sigma_h |a_{ik}|$$

$r = 1, 2, \ldots, m;\ s = m+1, m+2, \ldots, 2m;\ h = m+1, \ldots, 2m;$
$i = 1, 2, \ldots, m-1, h;\ k = m+1, \ldots, h-1, m, h+1, \ldots, 2m.$

Let A have independent elements. Let $|a_{rs}| = a$. The operation

$$\sum_{j=m}^{2m} a_{hj} \frac{\partial}{\partial a_{mj}} \left[\sum_{l=1}^{m} a_{lm} \frac{\partial a}{\partial a_{lh}} \right]$$

---

[1] Kowalewski: Einführung in die Determinantentheorie, p. 124.

[2] Kronecker: J. reine angew. Math. Vol. 72 (1870) pp. 152–175.

[3] Jacobi: J. reine angew. Math. Vol. 2 (1827) pp. 347–357.

[4] The last three proofs are by G. A. Bliss: Ann. of Math. II Vol. 16 (1914) pp. 43–44.

[5] Kronecker: S.-B. preuß. Akad. Wiss. 1882 II pp. 821–844 — Werke Vol. II pp. 389–397.

first replaces the elements of column h by elements with the same first subscript but with the second subscript m, and then replaces the elements of the last row of the new determinant by elements with the same second subscript but with the first subscript h. Consequently

$$\sum_{h=m+1}^{2m} \sum_{j=m}^{2m} a_{hj} \frac{\partial}{\partial a_{mj}} \sum_{l=1}^{m} a_{lm} \frac{\partial a}{\partial a_{lh}} = \sum_{h} |a_{ik}| .$$

The left member is equal to

$$\sum_{h=m+1}^{2m} \sum_{j=m+1}^{2m} a_{hj} \left[\sum_{l=1}^{m} a_{lm} \frac{\partial^2 a}{\partial a_{mj} \partial a_{lh}} \right] + \sum_{h=m+1}^{2m} a_{hm} \frac{\partial a}{\partial a_{mh}} .$$

Since $\frac{\partial a}{\partial a_{mm}} = 0$, $j \neq m$ in the second summation. For $h=j$, $\frac{\partial^2 a}{\partial a_{mj} \partial a_{lh}} = 0$. Now consider A to be symmetric. The coefficient of a_{hj}, $h \neq j$, is

$$\sum_{l=1}^{m} a_{lm} \left(\frac{\partial^2 a}{\partial a_{mj} \partial a_{lh}} + \frac{\partial^2 a}{\partial a_{mh} \partial a_{lj}} \right).$$

Since the interchange of columns h and j does not alter this expression, it vanishes. Hence the entire left member reduces to[1]

$$\sum_{h=m+1}^{2m} a_{mh} \frac{\partial a}{\partial a_{mh}} = a .$$

A formal proof of KRONECKER's identities was given by L. SCHENDEL[2]. H. S. WHITE[3] proved them from known identities in algebraic invariant theory. R. MEHMKE[4] stated that they are implied in GRASSMANN's "Linealen Ausdehnungslehre" 1862, p. 131.

C. RUNGE[5] proved that all linear relations among the minors of a symmetric matrix are linearly dependent upon those of KRONECKER, and found linearly independent systems. He also proved that no such relations exist for skew matrices. His results are significant in connection with the generalizations of KRONECKER's identities by MUIR[6], and HELEN BARTON[7].

METZLER[8] gave some complicated identities among the minors of a matrix which include KRONECKER's in the symmetric case. A very complete account of the identities among the minors of a matrix was recently given by R. A. BEAVER[9].

[1] STOUFFER, E. B.: Proc. Nat. Acad. Sci. U.S.A. Vol. 12 (1926) pp. 63—64.
[2] SCHENDEL, L.: Z. Math. Physik Vol. 32 (1887) pp. 119—120.
[3] WHITE, H. S.: Bull. Amer. Math. Soc. II Vol. 2 (1896) pp. 136—138.
[4] MEHMKE, R.: Math. Ann Vol. 26 (1886) pp. 209—210.
[5] RUNGE, C.: J. reine angew. Math. Vol. 93 (1882) pp. 319—327.
[6] MUIR: Philos. Mag. Vol. 3 (1902) pp. 410—416.
[7] BARTON, HELEN: Proc. Nat. Acad. Sci. U.S.A. Vol. 12 (1926) pp. 393—396.
[8] METZLER: Trans. Amer. Math. Soc. Vol. 2 (1901) pp. 395—403.
[9] BEAVER, R. A.: Amer. Math. Monthly Vol. 39 (1932) pp. 266—276.

MacMahon[1], and Muir[2] showed that there are just $n^2 - n + 1$ independent principal minors of a matrix. Stouffer[3] showed that (A_i), (A_{ij}), (A_{1ij}) constitute such a system, the subscripts indicating deleted rows and columns. Stouffer[4] gave a method for expressing the determinant of a matrix of any order in terms of not more than 14 of its principal minors of lower order, and later[5] showed that the determinant of a matrix of order n is a function of four minors of order $n-1$ and one of order $n-2$. A general theorem on the expression of a determinant in terms of its subdeterminants was proved by W. W. Flexner[6].

10. Reducibility. *Theorem 10.1. If A is either a general matrix or a general symmetric matrix, there is no identity*

$$d(A) = f(a_{ij})\, g(a_{ij}),$$

where f and g are polynomials in the elements of A neither of which is a constant.

Suppose the elements of A independent and $d(A) \equiv f \cdot g$. Since $d(A)$ is linear in every element, if a_{11} occurs in f it does not occur in g. No term of $d(A)$ contains $a_{11} a_{r1}$, hence g is of degree 0 in every a_{r1}. Since no term of $d(A)$ contains $a_{rs} a_{r1}$, g is of degree 0 in every a_{rs} and is therefore constant. Only a slight modification is required to extend the proof to the general symmetric matrix[7].

This is a special case of the theorem that the determinant of an irreducible group matrix is an irreducible function of the variables[8].

Theorem 10.2. The determinant of the general skew matrix of even order is the square of a rational function of its elements.

The theorem is obvious for $n = 2$, and the proof follows by induction. $A^{1,\ldots,n-1}_{1,\ldots,n-1}$ is skew of odd order and hence its rank is $n-2$ at most. (Theorem 8.6.) Let $A^{(n)}_{ij}$ be the cofactor of a_{ij} in $A^{1,\ldots,n-1}_{1,\ldots,n-1}$. As in the proof of Theorem 8.5,

$$A^{(n)}_{ij} = m\, p_i\, p_j,$$

where p_i is a rational function of the elements of A. By assumption $A^{(n)}_{ii}$ is the square of a rational function of the a_{ij}, so the same must be true of m, which may be taken as 1. By the Laplace development

$$d(A) = \Sigma_{i,j} A^{(n)}_{ij}\, a_{in}\, a_{jn} = (\Sigma_i\, p_i\, a_{in})^2\,.$$

The following more general theorem was proved by E. Zylinski[9]. Let $\bar{A}$ be a matrix obtained from $A = (a_{rs})$ by replacing a certain

---

[1] MacMahon: Philos. Trans. Roy. Soc. London Vol. 185 (1893) pp. 111—160.
[2] Muir: Philos. Mag. V Vol. 38 (1894) pp. 537—541.
[3] Stouffer: Trans. Amer. Math. Soc. Vol. 26 (1924) pp. 356—368.
[4] Stouffer: Amer. Math. Monthly Vol. 35 (1928) pp. 18—21.
[5] Stouffer: Amer. Math. Monthly Vol. 39 (1932) pp. 165—166.
[6] Flexner, W. W.: Ann. of Math. II Vol. 29 (1927) pp. 373—376.
[7] Bôcher: Introduction to higher algebra, p. 177. Macmillan 1907.
[8] Dickson: Modern algebraic theories, p. 259. Chicago 1926.
[9] Zylinski, E.: Bull. int. Acad. Polon. Sci. 1921 pp. 101—104.

number of elements by 0. Let A^* be a matrix obtained from A by replacing by 0's those of the elements which do not actually figure in the development of $d(\bar{A})$. A necessary and sufficient condition for the irreducibility of $d(A^*)$ is as follows: Starting from one element a_{rs} of A^*, one can reach each of the other elements (not 0) by a closed polygon each side of which is limited by two elements a_{rs} belonging to the same row or to the same column of A^*.

L. GEGENBAUER[1] proved that if the rows of a matrix are cyclic permutations of the first, in the field of the roots of unity the determinant is a product of linear factors. W. BURNSIDE[2] generalized this to the case where the successive rows proceed from the first by the permutations of an abelian group of order n.

FROBENIUS[3] proved that if the elements of the matrix X are independent variables, and those of the matrix Y linear functions of these variables, and if $d(X) = c\,d(Y), c \neq 0$, then either $Y = AXB$ or $Y = AX^{\mathrm{T}}B$ where A and B are constant matrices. For $n > 1$ only one of these relations can hold; A and B are unique up to scalar factors.

I. SCHUR[4] generalized the above theorem as follows: Let X be an array of m rows and n columns whose mn elements are independent variables. Let Y be an array of the same type whose mn elements y_{rs} are linear homogeneous functions of the elements x_{rs} of X. If for a fixed r, $2 \leqq r \leqq m$, $2 \leqq r \leqq n$, it is known that the $N = \binom{m}{r}\binom{n}{r}$ minors of order r of Y are linearly independent linear homogeneous functions of the N minors of X, then for $m \neq n$ the array Y is of the form AXB where A and B are non-singular constant square arrays of orders m and n respectively. If $m = n$, either $Y = AXB$ or $Y = AX^{\mathrm{T}}B$.

11. Arrays and determinants of higher dimension. The notion of extending the theory of determinants to cubic arrays and arrays of n dimensions has occurred to many writers, beginning with CAYLEY[5]. L. GEGENBAUER, M. LECAT and L. H. RICE, among others, have written extensively in that field. LECAT[6] gave a chronological list of 50 papers, and in 1926 brought the list up to date[7]. An excellent exposition was recently given by RICE[8]. The subject seems to lie much closer to tensor

[1] GEGENBAUER, L.: S.-B. Akad. Wiss. Wien (I, 2) Vol. 82 (1880—81) pp. 938 to 942.

[2] BURNSIDE, W.: Mess. Math. II Vol. 23 (1894) pp. 112—114.

[3] FROBENIUS: S.-B. preuß. Akad. Wiss. 1897 pp. 994—1015.

[4] SCHUR, I.: S.-B. preuß. Akad. Wiss. 1925 pp. 454—463.

[5] CAYLEY: Trans. Cambr. Philos. Soc. Vol. 8 (1843) pp. 1—16 — Coll. Works Vol. I pp. 75—79.

[6] LECAT, M.: Abrége de la théorie des déterminants à n dimensions. Gand. Hoste 1911.

[7] LECAT, M.: Ann. Soc. Sci. Bruxelles Vol. 46 (1926) pp. 1—39.

[8] RICE: J. Math. Physics, Massachusetts Inst. Technol. Vol. 9 (1930) pp. 47 to 71.

analysis than to the theory of matrices, as was clearly shown by C. M. CRAMLET[1].

A matrix is often considered as a linear vector function. J. D. BARTER[2] investigated homogeneous vector functions of degree p. He states that no analogue for characteristic root exists for the generalized matrix.

No success has as yet been attained in extending to higher dimensions the concept of matrix in the sense of hypercomplex number in which it is used in this book. Since every associative system can be represented in terms of two-dimensional matrices, this lack of success is not surprising. But n-dimensional arrays have received some attention in connection with multilinear forms and tensor analysis. Their importance lies in the various generalizations of rank which can be applied to them[3].

12. Matrices in non-commutative systems. Determinants of matrices whose elements are quaternions were discussed by HAMILTON[4]. J. M. PIERCE[5] defined a determinant of a matrix of quaternions by writing the elements of each term in the order of their column indices. J. BRILL[6] represented biquaternions and triquaternions as matrices with quaternion elements. E. STUDY[7] gave a brief discussion of matrices of quaternions.

Matrices whose elements belong to a division algebra or non-commutative field attain considerable importance from the theorem of J. H. M. WEDDERBURN[8] that every simple algebra can be represented as a total matric algebra whose elements belong to a division algebra[9].

Determinants of matrices over a division algebra in connection with the solution of systems of linear equations were considered by A. R. RICHARDSON[10], A. HEYTING[11], and O. ORE[12]. The latter defines the right-hand determinant by

$$|a_{rs}\| = a_{11}A_1^{(1)} + a_{21}A_1^{(2)} + \cdots + a_{n1}A_1^{(n)},$$

where the $A_1^{(j)}$ are a set of solutions of the homogeneous left-hand system

$$\sum_{i=1}^{n} a_{ij}A_1^{(j)} = 0. \qquad (j = 2, \ldots, n)$$

---

1 CRAMLET, C. M.: Amer. J. Math. Vol. 49 (1927) pp. 89–96.

2 BARTER, J. D.: Univ. California Publ. Math. Vol. 1 (1920) pp. 321–343.

3 HITCHCOCK, F. L.: J. Math. Physics, Massachusetts Inst. Technol. Vol. 7 (1927) pp. 39–85. — RICE, L. H.: Ibid. 1928 pp. 93–96.

4 HAMILTON: Elements of quaternions (Appendix). London 1889.

5 PIERCE, J. M.: Bull. Amer. Math. Soc. II Vol. 5 (1899) pp. 335–337.

6 BRILL, J.: Proc. London Math. Soc. II Vol. 4 (1906) pp. 124–130.

7 STUDY, E.: Acta math. Vol. 42 (1920) pp. 1–61.

8 WEDDERBURN, J. H. M.: Proc. London Math. Soc. II Vol. 6 (1908) p. 99.

9 See DICKSON: Algebren und ihre Zahlentheorie, p. 120. Zürich 1927.

10 RICHARDSON, A. R.: Mess. Math. Vol. 55 (1926) pp. 145–152 — Proc. London Math. Soc. II Vol. 28 (1928) pp. 395–420.

11 HEYTING, A.: Math. Ann. Vol. 98 (1927) pp. 465–490.

12 ORE, O.: Ann. of Math. II Vol. 32 (1931) pp. 463–477.

Two matrices are said to be right-equivalent if their right determinants vanish together. Two matrices A and B are right-equivalent if they differ by the interchange of two rows or of two columns; or if A is obtained from B by multiplying the elements of a column on the left, or the elements of a row on the right, with $k \neq 0$; or if A is obtained from B by adding one row to another row or one column to another column.

II. The characteristic equation.

13. The minimum equation. If A is a matrix of order n over a field $\mathfrak{F}$, the matrices $I, A, A^2, \ldots, A^{n^2}$ constitute $n^2 + 1$ sets of n^2 numbers each, and hence are linearly dependent in $\mathfrak{F}$. Thus A satisfies some equation

$$m(\lambda) = \lambda^{\mu} + m_1 \lambda^{\mu-1} + \cdots + m_{\mu} = 0$$

with coefficients in $\mathfrak{F}$ of minimum degree μ. We shall call μ the *index* of A. The index of a scalar matrix is 1. Every matrix except O has an index.

Theorem 13.1. If $f(A) = O$, $m(\lambda) \mid f(\lambda)$.[1]

Write $f(\lambda) = q(\lambda)\, m(\lambda) + r(\lambda)$ where $r(\lambda) = 0$ or else $r(\lambda)$ is of degree $< \mu$. Since $f(A) = m(A) = O$, $r(A) = O$. Since μ was minimal, $r(\lambda) = 0$.

Corollary 13.1. The minimum equation $m(\lambda) = 0$ is unique.

The constant term of the minimum equation will be called the *norm* of A, written $n(A)$.

14. The characteristic equation. The matrix obtained from $A = (a_{rs})$ by replacing every a_{rs} by the cofactor A_{sr} of a_{sr} is called the *adjoint* of A, written

$$A^{\mathrm{A}} = (A_{sr}).$$

If $d(A) \neq 0$, the matrix $A^{\mathrm{A}}/d(A)$ is called the *inverse* of A, written A^{I} or A^{-1}. By Theorem 7.8,

Theorem 14.1

$$A^{\mathrm{A}} A = A A^{\mathrm{A}} = I\, d(A),$$
$$A^{\mathrm{I}} A = A A^{\mathrm{I}} = I.$$

Theorem 14.2. If A satisfies an equation

$$p(\lambda) = C_0 \lambda^k + C_1 \lambda^{k-1} + \cdots + C_k = 0,$$

A and the C's being n-th order matrices with elements in a field $\mathfrak{F}$, then A satisfies the equation $d\,p(\lambda) = 0$ whose coefficients are in $\mathfrak{F}$.

Evidently $p(\lambda)$ can be considered as an n-th order matrix whose elements are polynomials in λ of degree $\leqq k$. Its adjoint

$$p^{\mathrm{A}}(\lambda) = D_0 \lambda^{k(n-1)} + D_1 \lambda^{k(n-1)-1} + \cdots + D_{k(n-1)}$$

---

[1] FROBENIUS: J. reine angew. Math. Vol. 84 (1878) p. 1–63.

has polynomial elements of degree $\leqq k(n-1)$. Then

$$d p(\lambda) = p_0 \lambda^{kn} + p_1 \lambda^{kn-1} + \cdots + p_{kn}$$

has coefficients in $\mathfrak{F}$. Since

$$p^{A}(\lambda) \cdot p(\lambda) = d p(\lambda) I,$$

$$\sum_{i=0}^{k(n-1)} D_{k(n-1)-i} \lambda^i \sum_{j=0}^{k} C_{k-j} \lambda^j = \sum_{h=0}^{kn} p_{kn-h} \lambda^h I,$$

$$\sum_{i=0}^{k(n-1)} D_{k(n-1)-i} \sum_{j=0}^{k} C_{k-j} \lambda^{i+j} = \sum_{h=0}^{kn} p_{kn-h} \lambda^h I.$$

This holds for λ indeterminate. By equating coefficients of the powers of λ, $kn+1$ equations are obtained. If the h-th equation is multiplied on the right by A^h and the results added, the sum may be written

$$\sum_{h=0}^{kn} p_{kn-h} A^h = \sum_{i=0}^{k(n-1)} D_{k(n-1)-i} \left[\sum_{j=0}^{k} C_{k-j} A^j \right] A^i = O,$$

or $d p(A) = O$.

This is a special case of a theorem of H. B. PHILLIPS[1] who proved that if $A_1, \ldots, A_k$ are n-th order matrices, and $B_1, \ldots, B_k$ are matrices commutative in pairs such that

$$A_1 B_1 + \cdots + A_k B_k = O,$$

and if $p(\beta_1, \ldots, \beta_k) = d(A_1 \beta_1 + \cdots + A_k \beta_k)$ where $\beta_1, \ldots, \beta_k$ are indeterminates of $\mathfrak{F}$, then

$$p(B_1, \ldots, B_k) = O.$$

The proof which we have given is similar to FROBENIUS'[2] proof for the linear case. FROBENIUS attributes the idea to PASCH[3].

Corollary 14.21. Every matrix satisfies its characteristic equation $d(\lambda I - A) = 0$.

This is the famous HAMILTON-CAYLEY theorem, established for quaternions by W. R. HAMILTON[4], and proved for $n = 2, 3$ by A. CAYLEY[5] who stated the theorem in the general case with the comment that it was not necessary to undertake its proof. Many proofs, more or less satisfactory, have been given[6]. A. R. FORSYTH[7] applied difference equations to the proof. A. BUCHHEIM made the proof essentially in the manner of FROBENIUS[8], stating that it was taken in concept from TAIT's Quaternions, p. 81.

---

[1] PHILLIPS, H. B.: Amer. J. Math. Vol. 41 (1919) pp. 266—278.

[2] FROBENIUS: S.-B. preuß. Akad. Wiss. 1896 p. 606.

[3] PASCH: Math. Ann. Vol. 38 (1891) pp. 24—49.

[4] HAMILTON, W. R.: Lectures on quaternions, p. 566. Dublin 1853.

[5] CAYLEY, A.: Philos. Trans. Roy. Soc. London Vol. 148 (1858) pp. 17—37.

[6] LAGUERRE, E.: J. École polytechn. Vol. 25 (1867) pp. 215—264 — Œuvres Vol. I pp. 228—233.

[7] FORSYTH, A. R.: Mess. Math. Vol. 13 (1884) pp. 139—142.

[8] FROBENIUS: Mess. Math. Vol. 13 (1884) pp. 62—66.

In connection with Theorem 14.2, it should be noted that BUCHHEIM[1] stated that the roots of the equation $|\lambda I - A| = 0$ are roots of $|p(\lambda)| = 0$.

Corollary 14.22. The index μ of A is $\leqq n$.

The name *characteristic equation* of A for the equation $|\lambda I - A| = 0$ is due to CAUCHY[2]. The left member is called the *characteristic function* of A.

Theorem 14.3. If

$$f(\lambda) = \lambda^n - t_1\lambda^{n-1} + t_2\lambda^{n-2} - \cdots \pm t_n = 0$$

is the characteristic equation of A, then t_i is the sum of all the principal i-rowed minors of A.

For the coefficient of λ^{n-i} in $|\lambda I - A| = 0$ is the sum of the determinants of all minors obtained by suppressing $n - i$ rows of $-A$ and the corresponding columns.

The first coefficient $t_1 = t(A) = a_{11} + \cdots + a_{nn}$ is called the *trace* of A. It is a scalar function of A second only in importance to $\pm t_n = d(A)$.

Theorem 14.4. Every equation of degree n with coefficients in $\mathfrak{F}$ is the characteristic equation of some matrix of order n with elements in $\mathfrak{F}$.

Let the given equation be

$$f(\lambda) = \lambda^n + b_1\lambda^{n-1} + \cdots + b_n = 0.$$

The matrix

$$B = \begin{Vmatrix} 0 & 1 & 0 & \cdots & 0 \\ 0 & 0 & 1 & \cdots & 0 \\ \cdot & \cdot & \cdot & \cdots & \cdot \\ -b_n & -b_{n-1} & -b_{n-2} & \cdots & -b_1 \end{Vmatrix}$$

has $f(\lambda) = 0$ as its characteristic equation. For if in the matrix $\lambda I - B$ one multiplies the last column by λ and adds to the $(n-1)$-th, multiplies the $(n-1)$-th column of the new matrix by λ and adds to the $(n-2)$-th and so on, there results

$$\begin{Vmatrix} 0 & -1 & 0 & 0 \\ 0 & 0 & -1 & 0 \\ \cdot & \cdot & \cdot & \cdot \\ f(\lambda) & * & * & * \end{Vmatrix}.$$

By the LAPLACE development, the determinant of this matrix is seen to be $\pm f(\lambda)$.[3]

[1] BUCHHEIM: Proc. London Math. Soc. Vol. 16 (1884) pp. 63–82.

[2] CAUCHY: Exercises d'analyse et de physique mathématique Vol. 1 (1840) p. 53 — Œuvres II Vol. 11 pp. 75–133.

[3] GÜNTHER, S.: Z. Math. Vol. 21 (1876) pp. 178–191. — LAISANT, C. A.: Bull. Soc. Math. France Vol. 17 (1889) pp. 104–107. — RADOS, G.: Math. Ann. Vol. 48 (1897) pp. 417–424.

A. LOEWY[1] called B (or its negative) a "Begleitmatrix". We shall take the liberty of calling it the *companion matrix* of the equation $f(\lambda) = 0$.

If $f(\lambda)$ is completely reducible in $\mathfrak{F}$, say

$$f(\lambda) = (\lambda - \lambda_1)^{l_1} \dots (\lambda - \lambda_k)^{l_k},$$

where the λ_i are distinct, let

$$J = \begin{Vmatrix} J_1 & 0 & \dots & 0 \\ 0 & J_2 & \dots & 0 \\ \cdot & \cdot & \cdot & \cdot \\ 0 & 0 & \dots & J_k \end{Vmatrix},$$

where

$$J_i = \begin{Vmatrix} \lambda_i & 1 & 0 & \dots & 0 \\ 0 & \lambda_i & 1 & \dots & 0 \\ \cdot & \cdot & \cdot & \cdot & \cdot \\ 0 & 0 & 0 & \dots & \lambda_i \end{Vmatrix}$$

is of order l_i. It is evident that $d(\lambda I - J) = f(\lambda)$. This matrix J will be called a JORDAN *matrix*[2].

15. Determination of the minimum equation. *Theorem 15.1. Let $f(\lambda) = 0$ be the characteristic equation of A, and let $h(\lambda)$ be the greatest common divisor of the $(n-1)$-rowed minors of $\lambda I - A$. Then*

$$g(\lambda) = f(\lambda)/h(\lambda) = 0$$

is the minimum equation of A.[3]

Let $\lambda I - A = C(\lambda)$. Then $C^A(\lambda) = h(\lambda) K(\lambda)$ where the elements of $K(\lambda)$ are relatively prime. Since $f(\lambda) I = C(\lambda) C^A(\lambda)$,

$$g(\lambda)\, h(\lambda)\, I = C(\lambda)\, h(\lambda)\, K(\lambda),$$

and, since $h(\lambda) \not\equiv 0$,

$$g(\lambda)\, I = C(\lambda)\, K(\lambda). \tag{15.1}$$

Since $C(A) = 0$, $g(A) = 0$ and $m(\lambda) \mid g(\lambda)$ where $m(\lambda) = 0$ is the minimum equation of A.

From $m(\lambda) - m(\mu) = (\lambda - \mu)\, k(\lambda, \mu)$, we obtain

$$m(\lambda)\, I - m(A) = m(\lambda I) - m(A) = C(\lambda)\, k(\lambda I, A).$$

But $m(A) = 0$. Hence

$$C^A(\lambda)\, m(\lambda) = f(\lambda)\, k(\lambda I, A),$$
$$h(\lambda)\, K(\lambda)\, m(\lambda) = h(\lambda)\, g(\lambda)\, k(\lambda I, A).$$

Since $h(\lambda) \not\equiv 0$, it may be canceled. Since $g(\lambda)$ divides every element of $K(\lambda) m(\lambda)$, and the elements of $K(\lambda)$ are relatively prime[4], $g(\lambda) \mid m(\lambda)$.

---

[1] LOEWY, A.: S.-B. Heidelberg. Akad. Wiss. Vol. 5 (1918) p. 3 — Math. Z. Vol. 7 (1920) pp. 58—125.

[2] JORDAN, C.: Traité des Substitutions et des Équations Algébriques, Livre 2 pp. 88—249. Paris 1870.

[3] FROBENIUS: J. reine angew. Math. Vol. 84 (1878) pp. 1—63.

[4] Proof by O. PERRON: Math. Ann. Vol. 64 (1906) pp. 248—263.

Corollary 15.1. If B is the companion matrix (or J the JORDAN *matrix) of* $f(\lambda) = 0$, *then* $f(\lambda) = 0$ *is the minimum equation of* B *(or* J*).*

Since $B_{n1} = 1$, the minimum equation of B is likewise its characteristic equation.

By crosing out of $\lambda I - J$ the first column and last row of $\lambda I_i - J_i$, a minor of order $n - 1$ is obtained whose determinant is free of $\lambda - \lambda_i$. Hence the g.c.d. of the $(n-1)$-rowed minor determinants of $\lambda I - J$ is 1, and the minimum equation of J is likewise its characteristic equation.

Theorem 15.2. The distinct factors of the characteristic function $f(\lambda)$ *of* A *which are irreducible in* $\mathfrak{F}$ *coincide with the distinct irreducible factors of the minimum function* $m(\lambda)$.

As in the proof of Theorem 15.1,

$$f(\lambda) = m(\lambda)\, h(\lambda).$$

Hence every root of $m(\lambda) = 0$ is a root of $f(\lambda) = 0$, so every irreducible factor of $m(\lambda)$ divides $f(\lambda)$. From (15.1)

$$C(\lambda)\, K(\lambda) = m(\lambda)\, I.$$

Taking determinants,

$$f(\lambda)\, dK(\lambda) = [m(\lambda)]^n.$$

Hence every root of $f(\lambda) = 0$ is a root of $m(\lambda) = 0$, and every irreducible factor of $f(\lambda)$ divides $m(\lambda)$.[1]

Corollary 15.2. If either $d(A)$ *or* $n(A)$ *vanishes, both vanish.*

Hence a matrix A is singular or non-singular according as $n(A) = 0$ or $n(A) \neq 0$.

A matrix is called *derogatory*[2] if its index μ is less than n.

Theorem 15.3. If A *is non-singular of order* n *and index* μ, A^{I} *is a polynomial in* A *of degree* $\mu - 1$.[3]

Let the minimum equation of A be

$$m(\lambda) = \lambda^{\mu} + m_1 \lambda^{\mu-1} + \cdots + m_{\mu-1}\lambda + n(A) = 0.$$

Set $B = -(A^{\mu-1} + m_1 A^{\mu-2} + \cdots + m_{\mu-1} I)/n(A)$.

Then $AB = BA = I$. But $AA^{\mathrm{I}} = A^{\mathrm{I}}A = I$, so $AB = AA^{\mathrm{I}}$, $A^{\mathrm{I}}AB = A^{\mathrm{I}}AA^{\mathrm{I}}$, $B = A^{\mathrm{I}}$.

Theorem 15.4. If A *is singular of order* n *and index* μ, *there exists a matrix* B *expressible as a polynomial in* A *of degree* $\mu - 1$ *such that* $AB = BA = 0$.

For if $B = -(A^{\mu-1} + m_1 A^{\mu-2} + \cdots + m_{\mu-1} I)$, $AB = BA = n(A) I = 0$.

Corollary 15.4. Every matrix is either non-singular or a divisor of zero.

Theorem 15.5. If $m(\lambda) = 0$ *is the minimum equation of* A, *then* $df(A) = 0$ *if and only if* $m(\lambda)$ *and* $f(\lambda)$ *have a common factor of degree* $\geqq 1$.[4]

---

[1] See L. E. DICKSON: Algebren und ihre Zahlentheorie, p. 21. Zürich 1927.

[2] SYLVESTER: C. R. Acad. Sci., Paris Vol. 98 I (1884) pp. 471—475.

[3] LAGUERRE: J. École polytechn. Vol. 25 (1867) pp. 215—264.

[4] HENSEL, K.: J. reine angew. Math. Vol. 127 (1904) pp. 116—166.

Suppose $m(\lambda)=h(\lambda)k(\lambda)$, $f(\lambda)=h(\lambda)\,l(\lambda)$, $h(\lambda)$ of degree $\geqq 1$. Then $k(\lambda)$ is of degree $<\mu$ and $k(A)\neq O$. Then

$$m(A)=O=h(A)\,k(A),$$

and $h(A)$ is singular. Then $f(A)$ is also singular.

Let $h(\lambda)$ be a g.c.d. of $m(\lambda)$ and $f(\lambda)$ so that

$$h(\lambda)=p(\lambda)\,m(\lambda)+q(\lambda)\,f(\lambda).$$

Then

$$h(A)=q(A)\,f(A).$$

If $f(A)$ is singular, $h(A)$ is also singular, and $h(\lambda)$ is not a constant.

Theorem 15.6. Let $r(x,y,z,\ldots)=\dfrac{p(x,y,z,\ldots)}{q(x,y,z,\ldots)}$ *be a rational function of the scalar indeterminates* $x,y,z,\ldots$, *where* p *and* q *are polynomials. Let* A *of index* λ, B *of index* μ, C *of index* $\nu,\ldots$ *be commutative matrices such that* $q(A,B,C,\ldots)$ *is non-singular. Then* $r(A,B,C,\ldots)$ *is uniquely defined, and can be represented as a polynomial of degree* $<\lambda$ *in* A, *of degree* $<\mu$ *in* B, *of degree* $<\nu$ *in* $C,\ldots$

A polynomial in $A,B,C,\ldots$ is built up by the operations of addition, multiplication and scalar multiplication. Since $A,B,C,\ldots$ are commutative, the result at each step is unique. Since $q(A,B,C,\ldots)$ is non-singular, q^{I} exists and is a polynomial in q, and therefore in $A,B,C,\ldots$ Hence $pq^{\mathrm{I}}=q^{\mathrm{I}}p$, and r is uniquely defined. By using the minimum equations of $A,B,C,\ldots$, the degree of r may be reduced as stated.

16. Characteristic roots. Let A be a matrix of order n with elements in a field $\mathfrak{F}$, and let $f(\lambda)=|\lambda I-A|$ be its characteristic function. Let $\mathfrak{F}'$ be an extension of $\mathfrak{F}$ in which $f(\lambda)=0$ is completely reducible. The n roots in $\mathfrak{F}'$ of $f(\lambda)=0$ are called the *characteristic roots*[1].

Theorem 16.1. Let $A,B,C,\ldots$ *be commutative matrices, and let* $f(x,y,z,\ldots)$ *be any rational function. The characteristic roots* $a_1,\ldots,a_n$ *of* A, $b_1,\ldots,b_n$ *of* B *etc. can be so ordered that the characteristic roots of* $f(A,B,C,\ldots)$ *are* $f(a_1,b_1,c_1,\ldots),\ldots,f(a_n,b_n,c_n,\ldots)$. *This ordering is the same for every function* f.[2]

Let

$$\alpha(\omega)=\prod_{i=1}^{\lambda}(\omega-a_i),\qquad \beta(\omega)=\prod_{j=1}^{\mu}(\omega-b_j),\qquad \gamma(\omega)=\prod_{k=1}^{\nu}(\omega-c_k),\quad\ldots$$

be the minimum equations of $A,B,C,\ldots$ respectively. Write $f(x,y,z,\ldots)-f(a_i,b_j,c_k,\ldots)=[f(x,b_j,c_k,\ldots)-f(a_i,b_j,c_k,\ldots)]+[f(x,y,c_k,\ldots)-f(x,b_j,c_k,\ldots)]+\cdots=(x-a_i)f_{ijk\ldots}+(y-b_j)g_{ijk\ldots}+(z-c_k)h_{ijk\ldots}$. Therefore

$$\prod_{i,j,k,\ldots}[f(x,y,z,\ldots)-f(a_i,b_j,c_k,\ldots)]=K\alpha(x)+L\beta(y)+M\gamma(z)+\cdots,$$

$$i=1,\ldots,\lambda,\quad j=1,\ldots,\mu,\quad k=1,\ldots,\nu,\ \ldots$$

[1] Or *latent roots*, SYLVESTER.

[2] FROBENIUS: S.-B. preuß. Akad. Wiss. 1896 I pp. 601–614.

Since $A, B, C, \ldots$ are commutative, these two polynomials give equal matrices when $x = A, y = B, z = C, \ldots$ Hence

$$\prod_{i,j,k,\ldots} [\omega - f(a_i, b_j, c_k, \ldots)] = 0$$

is satisfied by the matrix $f(A, B, C, \ldots)$. Therefore the minimum function of $f(A, B, C, \ldots)$, of degree $\sigma \leqq n$, is equal to a product of σ linear factors

$$\omega - f(a_i, b_j, c_k, \ldots).$$

By Theorem 15.2 these include all the distinct factors of the characteristic equation. The roots of $A, B, C, \ldots$ are ordered by calling the i-th root of $|\omega I - f| = 0$ $f(a_i, b_i, c_i, \ldots)$.

The above process can be carried through when f is a polynomial in x of degree $\lambda - 1$, in y of degree $\mu - 1$, in z of degree $\lambda - 1, \ldots$ with indeterminate coefficients, and an ordering of the roots of $A, B, C, \ldots$ so obtained. By Theorem 15.6 every rational function of $A, B, C, \ldots$ can be obtained by specializing the coefficients of this f. Hence the ordering is the same for all rational functions[1].

C. W. BORCHARDT[2] proved that the characteristic roots of A^n are the n-th powers of the characteristic roots of A. The special case for A symmetric was rediscovered by J. J. SYLVESTER[3]. W. SPOTTISWOODE[4] knew that the roots of A^I are the reciprocals of the roots of A. G. FROBENIUS[5] proved Theorem 16.1 first for a single matrix and later in the form above[6]. SYLVESTER rediscovered SPOTTISWOODE's[7] result and FROBENIUS'[8] theorem of 1878.

T. J. I'A. BROMWICH[9] noted that FROBENIUS' theorem need not hold when f is not a rational function.

Theorem 16.2. The characteristic roots of AB are the same as those of BA.[10]

By Theorems 14.3 and 7.9 the coefficient of $\pm\lambda^{n-i}$ in the characteristic equation of AB is

$$t_i = \sum A_{h_1, \ldots, h_i}^{r_1, \ldots, r_i} B_{r_1, \ldots, r_i}^{h_1, \ldots, h_i},$$

---

1 Proof by I. SCHUR: S.-B. preuß. Akad. Wiss. 1902 I pp. 120—125.

2 BORCHARDT, C. W.: J. reine angew. Math. Vol. 30 (1846) pp. 38—46 — J. Math. pures appl. I Vol. 12 (1847) pp. 50—67.

3 SYLVESTER, J. J.: Nouv. Ann. math. Vol. 11 (1852) pp. 439—440.

4 SPOTTISWOODE, W.: J. reine angew. Math. Vol. 51 (1856) pp. 209—271 and 328—381.

5 FROBENIUS, G.: J. reine angew. Math. Vol. 84 (1878) pp. 1—63.

6 FROBENIUS, G.: l. c.

7 SPOTTISWOODE: C. R. Acad. Sci., Paris Vol. 94 (1882) pp. 55—59.

8 FROBENIUS, G.: Philos. Mag. V Vol. 16 (1883) pp. 267—269.

9 BROMWICH, T. J. I'A.: Proc. Cambridge Philos. Soc. Vol. 11 (1901) pp. 75 to 89.

10 SYLVESTER: Philos. Mag. V Vol. 16 (1883) pp. 267—269.

where both $(r_1, \ldots, r_i)$ and $(h_1, \ldots, h_i)$ range over all $\binom{n}{i}$ selections of i distinct numbers of the set $1, 2, \ldots, n$ without regard for order. The theorem follows upon interchanging the summation indices.

SYLVESTER stated the theorem without proof. A proof based upon considerations of continuity was given by H. S. THURSTON[1]. A proof is sketched by H. W. TURNBULL and A. C. AITKEN[2].

A. CHÂTELET[3] noted that the theorem of FROBENIUS affords an easy method of applying a TSCHIRNHAUSEN transformation. Let it be required to find the equation whose roots are the rational function φ of the roots of $f(x) = 0$. Let B be the companion matrix of $f(x) = 0$. Then $|yI - \varphi(B)| = 0$ is the required equation.

If $f(x) = 0$ has algebraic integral roots, the elements of B are rational integers. If φ is a polynomial with rational integral coefficients, $\varphi(B)$ has rational integral elements, so that the roots of $|yI - \varphi(B)| = 0$ are algebraic integers. This is a short proof that a polynomial function with rational integral coefficients of an integral algebraic number is integral.

$A^\mu = I$ if and only if the characteristic roots of A are roots of unity[4].

If in the series $A, A^2, A^3, \ldots$ the same matrix appears more than once, then the system contains I. If ν is the minimum integer for which $A^\nu = I$, then the only different matrices in the set are $I, A, \ldots, A^{\nu-1}$.[5]

A. RANUM[6] gave a method for expressing all the powers of A as linear combinations of $I, A, \ldots, A^{n-1}$.

If for some ν $N^\nu = O$, then $d(N + A) = d(A)$.[7]

17. Conjugate sets. If the characteristic equation of the n-th order matrix A_1 is $f(\lambda) = 0$, a set of matrices $A_2, \ldots, A_n$ form with A_1 a *complete set of conjugates* if

1. They have the same characteristic equation,
2. They are commutative,
3. The elementary symmetric functions of these matrices are multiples of I by the corresponding elementary symmetric functions of the algebraic roots of $f(\lambda) = 0$.[8] If the elementary divisors of $\lambda I - A_1$ are linear, the conjugates always exist.

The product of the conjugates $A_2 A_3 \ldots A_n$ is $A_1{}^{\mathrm{A}}$. A. A. BENNETT[9]

[1] THURSTON, H. S.: Amer. Math. Monthly Vol. 38 (1931) pp. 322—324.

[2] TURNBULL, H. W., and A. C. AITKEN: An introduction to the theory of canonical matrices, p. 181. London 1932.

[3] CHÂTELET, A.: Ann. École norm. III Vol. 28 (1911) pp. 105—202.

[4] LIPSCHITZ, R.: Acta math. Vol. 10 (1887) pp. 137—144.

[5] KRONECKER, L.: S.-B. preuß. Akad. Wiss. 1890 pp. 1081—1088.

[6] RANUM, A.: Bull. Amer. Math. Soc. II Vol. 17 (1911) pp. 457—461.

[7] FROBENIUS: S.-B. preuß. Akad. Wiss. 1896 I pp. 601—614.

[8] TABER, H.: Amer. J. Math. Vol. 13 (1891) pp. 159—172.

[9] BENNETT, A. A.: Ann. of Math. II Vol. 23 (1923) pp. 91—96.

called $A_2 + \cdots + A_n$ the *adjoint-trace* of A_1, since its sum with A_1 is $t(A_1)I$.

P. FRANKLIN[1] modified the definition of conjugate set by the omission of the first condition, and showed that for every A_1 a set of generalized conjugate matrices exists, not always uniquely, however. Later[2] he gave a systematic method for finding all sets of conjugates for a given matrix.

J. WILLIAMSON[3] noted a particular matrix B such that $\omega^i B$ form a complete set of conjugate matrices in the sense of TABER, where ω is a primitive n-th root of unity.

T. A. PIERCE[4] proved that the field generated by the roots of any cyclic equation is isomorphic with the field generated by the matric roots of this equation.

18. Limits for the characteristic roots. In this paragraph $\mathfrak{F}$ is the complex field. Let A^C denote $(\bar{a}^{rs})$, obtained from A by replacing every element by its conjugate complex number.

The matrix H is *hermitian* if $H = H^{CT}$.[5] A real hermitian matrix is symmetric.

The matrix K is *skew-hermitian* if $K = -K^{CT}$.[6] A real skew-hermitian matrix is skew.

The matrix U is *unitary* if $U^I = U^{CT}$.[7] A real unitary matrix is *orthogonal.*

A matrix V is *involutory* if $V^I = V$, i.e., if $V^2 = I$.[8]

Theorem 18.1. If a matrix has two of the three properties in a set, it has all three.

1. *Real, orthogonal, unitary.*
2. *Symmetric, orthogonal, involutory.*
3. *Hermitian, unitary, involutory.*

The first of these is due to AUTONNE[9], the last two to H. HILTON[10].

Theorem 18.2. Every matrix is uniquely expressible $A = H + K$ where H is hermitian and K is skew-hermitian.

The proof is like that of theorem 5.3.

Theorem 18.3. If $A = H + K$ where H is hermitian and K is skew-hermitian, and if a, h, k, are the maxima of the absolute values of the

---

1 FRANKLIN, P.: Ann. of Math. II Vol. 23 (1923) pp. 97—100.

2 FRANKLIN, P.: J. Math. Physics, Massachusetts Inst. Technol. Vol. 10 (1932) pp. 289—314.

3 WILLIAMSON, J.: Amer. Math. Monthly Vol. 39 (1932) pp. 280—285.

4 PIERCE, T. A.: Bull. Amer. Math. Soc. Vol. 36 (1930) pp. 262—264.

5 HERMITE, C.: C. R. Acad. Sci., Paris Vol. 41 (1855) pp. 181—183.

6 LOEWY, A.: J. reine angew. Math. Vol. 122 (1900) pp. 53—72.

7 AUTONNE, L.: Rend. Circ. mat. Palermo Vol. 16 (1902) pp. 104—128.

8 VOSS, A.: Math. Ann. Vol. 13 (1878) pp. 320—374. — PRYM, F.: Abh. Ges. Wiss. Göttingen Vol. 38 (1892) pp. 1—42.

9 AUTONNE: Bull. Soc. Math. France Vol. 30 (1902) pp. 121—134.

10 HILTON, H.: Homogeneous linear substitutions. Oxford 1914.

elements of A, H, K, respectively, and if $\sigma = \alpha + \beta i$ is a characteristic root of A, then

$$|\sigma| \leqq na, \quad |\alpha| \leqq nh, \quad |\beta| \leqq nk.$$

Since $A - \sigma I$ is singular, there exists a vector $(x) = (x_1, \ldots, x_n)$ whose coördinates are not all 0 such that $(A - \sigma I)(x) = 0$. Hence

$$\sigma x_i = \Sigma_j a_{ij} x_j. \qquad (i, j = 1, 2, \ldots, n)$$

It then follows that

$$\sigma \Sigma_i x_i \bar{x}_i = \Sigma_{i,j} a_{ij} \bar{x}_i x_j,$$
$$\bar{\sigma} \Sigma_i x_i \bar{x}_i = \Sigma_{i,j} \bar{a}_{ji} \bar{x}_i x_j.$$

Hence, adding and subtracting,

$$\alpha \sum_i x_i \bar{x}_i = \sum_{i,j} \frac{a_{ij} + \bar{a}_{ji}}{2} \bar{x}_i x_j, \tag{18.1}$$

$$\beta \sum_i x_i \bar{x}_i = \sum_{i,j} \frac{a_{ij} - \bar{a}_{ji}}{2i} \bar{x}_i x_j. \tag{18.2}$$

Now

$$|\sigma| \sum x_i \bar{x}_i \leqq \sum_{i,j} |a_{ij}| \cdot |x_i| \cdot |x_j| \leqq a \sum |x_i| \cdot |x_j| = a \Big[\sum_i |x_i|\Big]^2.$$

From the inequality

$$(k_1 + \cdots + k_m)^2 \leqq m(k_1^2 + \cdots + k_m^2), \tag{18.3}$$

it follows that

$$[\sum |x_i|]^2 \leqq n \sum x_i \bar{x}_i, \qquad |\sigma| \leqq na.$$

Similarly[1]

$$|\alpha| \leqq nh, \qquad |\beta| \leqq nk.$$

Corollary 18.31. The characteristic roots of an hermitian matrix are all real.

For if A is hermitian, $K = O$, $k = 0$, and every $|\beta| = 0$.

The first proof that the roots of a real symmetric matrix are all real is due to A. CAUCHY[2]. Many later proofs have been given by JACOBI, SYLVESTER, BUCHHEIM etc. The extension to matrices such that $A^{\mathrm{T}} = A^{\mathrm{C}}$ was made by HERMITE[3] and resulted in such matrices being named for him.

Corollary 18.32. The characteristic roots of a skew-hermitian matrix are all pure imaginaries[4].

For in this case $h = 0$, $|\alpha| = 0$.

This theorem was proved for skew matrices by A. CLEBSCH[5] and later by WEIERSTRASS[6].

[1] Theorem and proof by A. HIRSCH: Acta math. Vol. 25 (1901) pp. 367–370.
[2] CAUCHY, A.: Anciens Exercises. 1829–1830 — Coll. Works II Vol. 9 pp. 174 to 195.
[3] HERMITE: C. R. Acad. Sci., Paris Vol. 41 (1855) pp. 181–183.
[4] SCORZA, G.: Corpi numerici e Algebre pp. 133–179. Messina 1921.
[5] CLEBSCH, A.: J. reine angew. Math. Vol. 62 (1863) pp. 232–245.
[6] WEIERSTRASS: S.-B. preuß. Akad. Wiss. 1879.

Theorem 18.4. If H in the last theorem is real, then

$$|\beta| \leqq \sqrt{\frac{n(n-1)}{2}}\, k.$$

From (18.2),

$$\beta \sum_i x_i \bar{x}_i = \sum_{i<j} \left[\frac{a_{ij} - \bar{a}_{ji}}{2}\right] \left[\frac{\bar{x}_i x_j - x_i \bar{x}_j}{i}\right],$$

$$|\beta| \sum_i x_i \bar{x}_i \leqq k \sum_{i<j} \frac{\bar{x}_i x_j - x_i \bar{x}_j}{i}.$$

Since $(\bar{x}_i x_j - x_i \bar{x}_j)/i$ are all real, the inequality (18.3) may be applied to give

$$\left[\sum_{i<j} \left|\frac{\bar{x}_i x_j - x_i \bar{x}_j}{i}\right|\right]^2 \leqq -\frac{n(n-1)}{2} \sum_{i<j} (\bar{x}_i x_j - x_i \bar{x}_j)^2$$

$$= \frac{n(n-1)}{2} \left\{\left[\sum x_i \bar{x}_i\right]^2 - \sum x_i^2 \cdot \sum \bar{x}_i^2\right\} < \frac{n(n-1)}{2} \left[\sum x_i \bar{x}_i\right]^2.$$

Hence $|\beta| \leqq \sqrt{\frac{n(n-1)}{2}}\, k$.[1]

The case when A is real was proved by I. BENDIXSON[2], and the paper of BENDIXSON seems to have inspired the work of HIRSCH.

Theorem 18.5. Let $\sigma = \alpha + i\beta$ be a characteristic root of $A = H + K$. Then $m \leqq \alpha \leqq M$ where m is the least and M the greatest of the characteristic roots of H.

Let the elements of the hermitian matrix H be denoted by h_{rs}. By (18.1) α is a value of the ratio

$$f(x) = \sum_{i,j} h_{ij} x_i \bar{x}_j \Big/ \sum_i x_i \bar{x}_i,$$

where the $x_1, \ldots, x_n$ are independent complex variables. The greatest and least values which $f(x)$ assumes are characteristic roots of H, for upon setting $x_j = u_j + i v_j$, the conditions

$$\frac{\partial f}{\partial u_j} = \frac{\partial f}{\partial v_j} = 0 \qquad (j = 1, 2, \ldots, n)$$

lead to the equation

$$\sum_{i,j} h_{ij} x_i \bar{x}_j - f \sum_i x_i \bar{x}_i = 0.$$

Hence m is the minimum value which $f(x)$ assumes and M is the maximum value. Then $m \leqq \alpha \leqq M$.[3]

BROMWICH[4] found an explicit expression for α and β from which HIRSCH's inequalities may be derived.

I. SCHUR[5] derived the inequalities by another method.

---

[1] This theorem and proof are by A. HIRSCH: l. c.

[2] BENDIXSON, I.: Acta math. Vol. 25 (1901) pp. 359—365.

[3] The real case by BENDIXSON, the complex case by HIRSCH: l. c.

[4] BROMWICH: Acta math. Vol. 30 (1906) pp. 295—304.

[5] SCHUR, I.: Math. Ann. Vol. 66 (1909) pp. 488—510.

E. T. BROWNE[1] proved that if λ is a characteristic root of A and if m is the least and M the greatest characteristic root of the hermitian matrix AA^{CT}, then $m \leqq \lambda\bar{\lambda} \leqq M$.

FROBENIUS[2] proved that if the elements of A are all real and positive, it has a characteristic root which is real, positive, simple, and greater in absolute value than any other characteristic root.

Theorem 18.6. The roots of the minimum equation of an hermitian matrix are distinct.

If the hermitian matrix H has the minimum equation

$$m(\lambda) = (\lambda - \lambda_1)^k h(\lambda),$$

where $k > 1$, define

$$m_1(\lambda) = (\lambda - \lambda_1)^{k-1} h(\lambda).$$

Then $m(\lambda) \mid [m_1(\lambda)]^2$. The matrix $H_1 = (h_{rs}) = m_1(H)$ is hermitian, and $H_1^2 = O$ so that $t(H_1^2) = 0$. But

$$t(H_1^2) = \sum_{i,j} h_{ij} h_{ji} = \sum_{i,j} h_{ij} \bar{h}_{ij},$$

so that $H_1 = O$ and $m_1(H) = O$. But $m_1(\lambda)$ is of lower degree than $m(\lambda)$. Hence the assumption $k > 1$ leads to a contradiction[3].

19. Characteristic roots of unitary matrices. *Theorem 19.1. The characteristic roots of a unitary matrix are of absolute value unity.*

If x is a characteristic root of the unitary matrix U, $x + 1/x$ is a root of the hermitian matrix $U + U^{I} = U + U^{CT}$, and $x - 1/x$ is a root of the skew-hermitian matrix $U - U^{I} = U - U^{CT}$ by Theorem 16.1. Hence by Corollaries 18.31 and 18.32,

$$x + 1/x = 2r, \qquad x - 1/x = 2is,$$

where r and s are real. That is,

$$x = r + is, \qquad 1/x = r - is,$$

whence $r^2 + s^2 = 1$.[4]

Another short proof was given by R. BRAUER[5]. If $U = (u_{rs})$ is unitary, $\Sigma_i u_{ri} \bar{u}_{ri} = 1$, so the elements of U are bounded in absolute value. The same is therefore true of the coefficients of the characteristic equation of U, and of its roots. But U^{I} and U^k are likewise unitary, and the characteristic roots of U^{I} are the reciprocals of those of U, the characteristic roots of U^k are the k-th powers of those of U.[6] Unless each characteristic root of U were of absolute value unity, some positive or negative power of it could be made arbitrarily large, thus leading to a contradiction.

---

[1] BROWNE, E. T.: Bull. Amer. Math. Soc. Vol. 34 (1928) pp. 363–368.
[2] FROBENIUS: S.-B. preuß. Akad. Wiss. 1908 pp. 471–476.
[3] WEDDERBURN, J. H. M.: Ann. of Math. II Vol. 27 (1926) pp. 245–248.
[4] ARAMATA, H.: Tôhoku Math. J. Vol. 28 (1927) p. 281.
[5] BRAUER, R.: Tôhoku Math. J. Vol. 30 (1928) p. 72.
[6] Theorem 16.1.

If U is unitary, and if $VV^{CT} = V^{CT}V$, and if ϱ is a root of $|U - \varrho V| = 0$, then $|t_1| \geqq 1/|\varrho| \geqq |t_2|$ where t_1 is the greatest in absolute value and t_2 the least in absolute value of the characteristic roots of V. Each characteristic root of a principal minor of a unitary matrix is $\leqq 1$ in absolute value[1].

Corollary 19.1. The complex characteristic roots of an orthogonal matrix occur in reciprocal pairs[2].

Since the characteristic equation of an orthogonal matrix has real coefficients, the complex roots occur in conjugate pairs, which by Theorem 19.1 are reciprocal pairs.

Other proofs of this theorem have been given by A. E. RAHUSEN[3] and A. COLUCCI[4]. In connection with this latter paper see also G. VITALI[5].

P. BURGATTI[6] proved that if $|A + xI| = 0$ has reciprocal roots, A is either orthogonal or involutory.

L. TOSCANO[7] proved that if the coefficients equidistant from the ends of the characteristic equation of A are equal, $|A + A^A + xI| = 0$ has all real roots, while[8] if these coefficients occur in pairs with opposite signs, $|A - A^A + xI| = 0$ has all real roots.

III. Associated Integral Matrices.

20. Matrices with elements in a principal ideal ring. A commutative ring without divisors of zero is called a *domain of integrity*. A domain of integrity with unit element 1 in which every pair of elements not both 0 has a greatest common divisor representable linearly in terms of the elements is called a *principal ideal ring*[9].

In a principal ideal ring $\mathfrak{P}$, a number which divides 1 is called a *unit*. The relation $a = ub$ where u is a unit is reciprocal, and two numbers a and b so related are called *associates*. A set of numbers of $\mathfrak{P}$ no two of which are associated but such that every number of $\mathfrak{P}$ is associated with one of them is called a *complete set of non-associates* in $\mathfrak{P}$. Thus the positive integers and 0 constitute a *complete set of non-associates* in the ring of rational integers.

A number a of $\mathfrak{P}$ neither 0 nor a unit is *prime* or *composite* according as $a = bc$ implies or does not imply that b or c is a unit.

1 LOEWY, A., and R. BRAUER: Tôhoku Math. J. Vol. 32 (1929) pp. 44–49.
2 BRIOSCHI, F.: J. Math. pures appl. Vol. 19 (1854) pp. 253–256.
3 RAHUSEN, A. E.: Wiskundige Opgaven Vol. 5 (1893) pp. 392–394.
4 COLUCCI, A.: Boll. Un. Mat. Ital. Vol. 6 (1927) pp. 258–260.
5 VITALI, G.: Boll. Un. Mat. Ital. Vol. 7 (1928) pp. 1–7.
6 BURGATTI, P.: Boll. Un. Mat. Ital. Vol. 7 (1928) pp. 65–69.
7 TOSCANO, L.: Atti Accad. naz. Lincei, Rend. VI Vol. 8 (1928) pp. 664–669.
8 TOSCANO, L.: Tôhoku Math. J. Vol. 32 (1929) pp. 27–31.
9 VAN DER WAERDEN: Moderne Algebra Vol. I pp. 39 and 60. Berlin: Julius Springer 1930.

An important instance of a principal ideal ring is a *euclidean ring* defined by the following properties:

1. Associated with every element a (except possibly 0) of the ring, there is a positive or 0 integer $s(a)$ called the *stathm* of a.

2. For every pair of numbers a and b, $b \neq 0$, there exist two numbers r and q such that $a = bq + r$, and either $r = 0$ or else $s(r) < s(b)$.

Thus for the GAUSSIAN complex numbers $a + ib$, a stathm is $a^2 + b^2$. For the polynomial domain $\mathfrak{F}(x)$ of a commutative field $\mathfrak{F}$, the degree of the polynomial serves as a stathm. For the trivial instance of a field, $s(a)$ may be taken as 1 for every a. A euclidean greatest common divisor process exists in every euclidean ring.

The euclidean ring $\mathfrak{E}$ is *proper* if it is not a field and if a stathm can be determined such that $s(ab) = s(a)\,s(b)$ for every a and b. Since $s(0 \cdot b) = s(0)\,s(b)$, either $s(b) = 1$ for every b so that division is always possible and $\mathfrak{E}$ is a field, or else $s(0) = 0$. Conversely let $s(a) = 0$. Either $a = 0$ or else for every b, $b = qa$, and $\mathfrak{E}$ is again a field.

Since $s(1 \cdot b) = s(1)\,s(b)$, either $s(b) = 0$ (and $\mathfrak{E}$ is a field) or $s(1) = 1$. Let u be a unit, and $uv = 1$. Then $s(u)\,s(v) = 1$ so that $s(u) = 1$. Conversely if $s(a) = 1$, a is a unit, for the stathm of the remainder upon dividing any number by a must be <1 and hence be 0. Thus *in a proper euclidean ring* $s(a) = 0$ *if and only if* $a = 0$, $s(u) = 1$ *if and only if* u *is a unit, and* $s(a) = s(b)$ *if and only if* a *and* b *are associated*[1].

Let $\mathfrak{M}$ constitute the set of all n-th order matrices whose elements belong to a principal ideal ring $\mathfrak{P}$ subject to the operations of addition and multiplication as in § 3. $\mathfrak{M}$ is a ring but not a principal ideal ring.

A matrix U of $\mathfrak{M}$ is called *unimodular* or a *unit matrix* if and only if there exists a matrix U' of $\mathfrak{M}$ such that $UU' = I$.

Theorem 20.1. U is a unit matrix if and only if $d(U)$ is a unit of $\mathfrak{P}$.

For $UU' = I$ implies $d(U)\,d(U') = 1$ so that $d(U)$ is a unit of $\mathfrak{P}$. Conversely if U is in $\mathfrak{M}$ and $d(U)$ is a unit of $\mathfrak{P}$, then U^{I} is in $\mathfrak{M}$ and serves as the U' of the definition.

Theorem 20.2. A matrix A of $\mathfrak{M}$ is a divisor of zero if and only if $d(A) = 0$.

For if A is in $\mathfrak{M}$, the matrix B of Theorem 15.4 (such that $AB = BA = 0$) is also in $\mathfrak{M}$.

A matrix A neither a divisor of zero nor a unit is called *prime* if every relation $A = BC$ implies that either B or C is a unit. Matrices neither divisors of zero nor units nor primes are called *composite*.

Theorem 20.3. A composite matrix can be expressed as a product of at most a finite number of primes.

For

$$A = A_1 A_2 A_3 \ldots,$$

[1] See J. H. M. WEDDERBURN: J. reine angew. Math. Vol. 167 (1931) pp. 129 to 141.

where no A_i is a unit matrix implies

$$d(A) = d(A_1)\,d(A_2)\,d(A_3)\,\ldots,$$

where no $d(A_i)$ is 0 or a unit. That is, each element in the sequence

$$d(A_1)\,d(A_2)\,d(A_3)\,\ldots, \quad d(A_2)\,d(A_3)\,\ldots, \quad d(A_3)\,\ldots$$

is a proper divisor of the preceding. Such a sequence can contain but a finite number of elements[1].

21. Construction of unimodular matrices. *Theorem 21.1. Let $a_1, a_2, \ldots, a_n$ be numbers of a principal ideal ring $\mathfrak{P}$ with greatest common divisor d_n. There exists a matrix of determinant d_n having $a_1, a_2, \ldots, a_n$ as its first row*[2].

The theorem is evidently true for $n = 2$. Suppose it true for $n - 1$, and let D_{n-1} be a matrix which has $a_1, a_2, \ldots, a_{n-1}$ as its first row, and whose determinant is the g. c. d. d_{n-1} of $a_1, a_2, \ldots, a_{n-1}$. Determine p and q so that $p d_{n-1} - q a_n = d_n$. Consider the matrix D_n obtained by bordering D_{n-1} on the right by $a_n, 0, \ldots, 0, 0$ and then below by $a_1 q/d_{n-1}, a_2 q/d_{n-1}, \ldots, a_{n-1} q/d_{n-1}, p$. Then $d(D_n) = d_n$.[3]

Other proofs have been given by K. WEIHRAUCH[4], BIANCHI[5], and H. HANCOCK[6].

HERMITE[7] proved more generally that a matrix can be found with a given last row $a_{n1}, a_{n2}, \ldots, a_{nn}$ and a determinant $k_1 a_{n1} + k_2 a_{n2} + \cdots + k_n a_{nn}$ where the k's are arbitrary.

The theorem has been generalized as follows. A $p \cdot n$ array of rational integers is given, $p < n$, and the g. c. d. of the p-rowed minor determinants is d. It is always possible to add $n - p$ rows of integers to this array so that the resulting square array shall be of determinant d. H. J. S. SMITH[8] and T. J. STIELTJES[9] proved this theorem and determined all possible borders. BLOCH[3] gave a short proof by induction.

22. Associated matrices. Two matrices A and B are called *left associates* if there exists a unit matrix U such that $A = UB$. The notation $A \stackrel{L}{=} B$ will be used to express this relationship, which possesses the prerequisites for an equality relationship, namely[10]:

---

[1] VAN DER WAERDEN: Moderne Algebra Vol. I p. 64. Berlin: Julius Springer 1930.

[2] For rational integers, $n = 3$, G. EISENSTEIN: J. reine angew. Math. Vol. 28 (1844) pp. 289—374. For any n, C. HERMITE: J. Math. pures appl. I Vol. 14 (1849) pp. 21—30.

[3] BLOCH, A.: Bull. Soc. Math. France Vol. 50 (1922) p. 100—110.

[4] WEIHRAUCH, K.: Z. Math. Physik Vol. 21 (1876) pp. 134—137.

[5] BIANCHI: Lezioni sulla Teoria dei Numeri Algebrici, pp. 1—7.

[6] HANCOCK, H.: Amer. Math. Monthly Vol. 31 (1924) pp. 161—162.

[7] HERMITE: J. reine angew. Math. Vol. 40 (1850) pp. 261—278.

[8] SMITH, H. J. S.: Philos. Trans. Roy. Soc. London Vol. 151 (1861—1862) pp. 293—326.

[9] STIELTJES, T. J.: Ann. Fac. Sci. Univ. Toulouse Vol. 4 (1890) pp. 1—103.

[10] ORE, O.: Bull. Amer. Math. Soc. Vol. 37 (1931) p. 538.

1. Determination. Either $A \overset{\mathrm{L}}{=} B$ or $A \overset{\mathrm{L}}{\neq} B$.
2. Reflexivity. $A \overset{\mathrm{L}}{=} A$.
3. Symmetry. $A \overset{\mathrm{L}}{=} B$ implies $B \overset{\mathrm{L}}{=} A$.
4. Transitivity. If $A \overset{\mathrm{L}}{=} B$ and $B \overset{\mathrm{L}}{=} C$, then $A \overset{\mathrm{L}}{=} C$.

A similar theory holds for the relation $A \overset{\mathrm{R}}{=} B$, or $A = BU$.

The following *elementary operations* upon the rows of a matrix can be accomplished by multiplying the matrix on the left by an *elementary matrix,* namely the matrix obtained by performing the elementary operation under consideration upon the identity matrix I.

1. The interchange of any two rows.
2. The multiplication of the elements of a row by a unit u of $\mathfrak{P}$.
3. The addition to the elements of any row of k times the corresponding elements of any other row, where k is in $\mathfrak{P}$.

Every elementary matrix is a unit matrix whose inverse is an elementary matrix of the same type. If B is obtainable from A by a finite number of elementary transformations, $B \overset{\mathrm{L}}{=} A$.[1]

Theorem 22.1. Every matrix A with elements in $\mathfrak{P}$ is the left associate of a matrix having 0's above the main diagonal, each diagonal element lying in a prescribed system of non-associates, and each element below the main diagonal lying in a prescribed residue system modulo the diagonal element above it.

A matrix of this type will be said to be in HERMITE's *normal form.*

Let $A = (a_{rs})$ have elements in $\mathfrak{P}$. Either every element of the last column is 0 or there is at least one non-zero element which by a permutation of the rows can be put into the (n, n)-position. Let d_n be a g.c.d. of the elements of the last column, and suppose that

$$b_1 a_{1n} + b_2 a_{2n} + \cdots + b_n a_{nn} = d_n .$$

By Theorem 21.1 there exists a unimodular matrix U having $b_1, b_2, \ldots, b_n$ as its last row. Then $A_1 = UA \overset{\mathrm{L}}{=} A$ has d_n in the (n, n)-position, where d_n divides every a_{in}. By subtracting a proper multiple of the last row from each of the other rows, a matrix A_1 is obtained whose last column has only 0's above the main diagonal.

In the $(n-1)$-th column either every element of the first $n-1$ rows is 0, or by a permutation of the first $n-1$ rows a non-zero element can be put into the $(n-1, n-1)$-position. Let d_{n-1} be a g.c.d. of $a_{1,n-1}, a_{2,n-1}, \ldots, a_{n-1,n-1}$ and let

$$b_1 a_{1,n-1} + b_2 a_{2,n-1} + \cdots + b_{n-1} a_{n-1,n-1} = d_{n-1} .$$

Let

$$U_1 = \begin{Vmatrix} u_{11} & u_{12} & \ldots & u_{1,n-1} & 0 \\ \cdot & \cdot & \cdot & \cdot & \cdot \\ b_1 & b_2 & \ldots & b_{n-1} & 0 \\ 0 & 0 & \ldots & 0 & 1 \end{Vmatrix}$$

[1] See L. KRONECKER: M.-B. preuß. Akad. Wiss. 1866 pp. 597—612.

be unimodular. Then $U_1 A_1$ has 0's above the main diagonal in the last column, d_{n-1} in the $(n-1, n-1)$-position, and each element above d_{n-1} divisible by d_{n-1}, so that these can be made equal to 0 by elementary transformations. The process is continued until a matrix is obtained which has only 0's above the main diagonal.

In order to make a_{ii} lie in any prescribed system of non-associates, it is at most necessary to multiply it by a unit. This is accomplished by an elementary transformation of the second type.

By subtracting a multiple of the $(n-1)$-th row from the n-th row, $a_{n,n-1}$ can be made to lie in any prescribed residue system modulo $a_{n-1,n-1}$. Similarly every element can be reduced modulo the diagonal element above it. It is understood that $a \equiv b \mod 0$ if and only if $a = b$.

Theorem 22.2. If $d(A) \neq 0$, the normal form of HERMITE *is unique.*

The proof will be made for $n = 3$, but the process is general. Suppose

$$\begin{Vmatrix} l_{11} & l_{12} & l_{13} \\ l_{21} & l_{22} & l_{23} \\ l_{31} & l_{32} & l_{33} \end{Vmatrix} \begin{Vmatrix} a_{11} & 0 & 0 \\ a_{21} & a_{22} & 0 \\ a_{31} & a_{32} & a_{33} \end{Vmatrix} = \begin{Vmatrix} b_{11} & 0 & 0 \\ b_{21} & b_{22} & 0 \\ b_{31} & b_{32} & b_{33} \end{Vmatrix}$$

where both $A = (a_{rs})$ and $B = (b_{rs})$ are in normal form and $L = (l_{rs})$ is unimodular.

$$b_{13} = l_{13} a_{33} = 0, \qquad b_{23} = l_{23} a_{33} = 0.$$

Since $d(A) \neq 0$, $a_{33} \neq 0$ and $l_{13} = l_{23} = 0$. Similarly every element of L above the main diagonal is 0.

Since $d(L)$ is a unit of $\mathfrak{P}$, l_{11}, l_{22}, l_{33} are all units of $\mathfrak{P}$. Then $l_{ii} a_{ii} = b_{ii}$, and since a_{ii} and b_{ii} belong to the same system of non-associates, $l_{ii} = 1$. Then

$$b_{32} = l_{32} a_{22} + a_{32}.$$

Since b_{32} and a_{32} lie in the same residue system modulo a_{22}, $l_{32} = 0$. Then

$$b_{31} = l_{31} a_{11} + a_{31},$$

and similarly $l_{31} = 0$. Thus $L = I$.

Corollary 22.2. Every non-singular matrix whose elements are rational integers is the left associate of a matrix whose diagonal elements a_{ii} are positive, $a_{ri} = 0$ for $r < i$, and $0 \leq a_{ri} < a_{ii}$ for $r > i$. This form is unique[1].

Theorem 22.3. If A has elements in a euclidean ring $\mathfrak{E}$, the reduction to normal form can be accomplished by elementary transformations.

Let $A = (a_{rs})$ have elements in $\mathfrak{E}$. Either every element of the last column is 0 or there is at least one non-zero element with minimum positive stathm which by an interchange of the rows can be put into the (n, n)-position. If a_{nn} does not divide some a_{in}, set

$$a_{in} = q\, a_{nn} + r. \qquad (s(r) < s(a_{nn}))$$

[1] HERMITE, C.: J. reine angew. Math. Vol. 41 (1851) pp. 191–216.

Then by an elementary transformation of the third type, a_{in} can be replaced by r. Again interchange rows if necessary so that the element in the (n, n)-position is of minimum stathm and proceed as before. Eventually a_{nn} will divide every a_{in}. The proof now proceeds as in Theorem 22.1.

Theorem 22.4. Every unimodular matrix with elements in a euclidean ring $\mathfrak{E}$ *is a product of a finite number of elementary matrices.*

Let U be unimodular. By Theorem 22.3 there exist elementary matrices E_i such that

$$E_1 E_2 \ldots E_k U = U',$$

where U' is unimodular and reduced. Since $d(U')$ is a unit of $\mathfrak{E}$, each diagonal element of U' is a unit of $\mathfrak{E}$, and may be taken to be 1. Since 0 constitutes a complete system of residues modulo 1, $U' = I$. Then

$$U = E_k^{\mathrm{I}} E_{k-1}^{\mathrm{I}} \ldots E_2^{\mathrm{I}} E_1^{\mathrm{I}},$$

where each E_i^{I} is an elementary matrix.

Corollary 22.4. Every unimodular matrix with elements in a euclidean ring having but a finite number of units is a product of a finite number of matrices of a certain finite set.

This was proved for rational integers by KRONECKER[1].

Theorem 22.5. Every unimodular matrix with elements in a euclidean ring $\mathfrak{E}$ *is a product of matrices of the types*

$$U_1 = \begin{Vmatrix} 0 & 0 & \ldots & 0 & 1 \\ 1 & 0 & \ldots & 0 & 0 \\ \cdot & \cdot & \cdots & \cdot & \cdot \\ 0 & 0 & \ldots & 0 & 0 \\ 0 & 0 & \ldots & 1 & 0 \end{Vmatrix}, \qquad U_2 = \begin{Vmatrix} 0 & 1 & \ldots & 0 & 0 \\ 1 & 0 & \ldots & 0 & 0 \\ \cdot & \cdot & \cdots & \cdot & \cdot \\ 0 & 0 & \ldots & 1 & 0 \\ 0 & 0 & \ldots & 0 & 1 \end{Vmatrix},$$

$$U_3(k) = \begin{Vmatrix} 1 & k & \ldots & 0 & 0 \\ 0 & 1 & \ldots & 0 & 0 \\ \cdot & \cdot & \cdots & \cdot & \cdot \\ 0 & 0 & \ldots & 1 & 0 \\ 0 & 0 & \ldots & 0 & 1 \end{Vmatrix}, \qquad U_4(\varepsilon) = \begin{Vmatrix} \varepsilon & 0 & \ldots & 0 & 0 \\ 0 & 1 & \ldots & 0 & 0 \\ \cdot & \cdot & \cdots & \cdot & \cdot \\ 0 & 0 & \ldots & 1 & 0 \\ 0 & 0 & \ldots & 0 & 1 \end{Vmatrix}.$$

The effect of U_1 is to permute the rows cyclicly. The effect of U_2 is to interchange the first two rows. By repeated use of these two operations any two rows of a matrix can be brought into positions one and two. Then by use of U_2 or U_3 or U_4 the desired elementary operation can be performed, and the rows then restored to their original positions.

This theorem is essentially that of A. KRAZER[2] who proved that

---

[1] KRONECKER: M.-B. preuß. Akad. Wiss. 1866 pp. 597—612.

[2] KRAZER, A.: Ann. Mat. pura appl. II Vol. 12 (1884) pp. 283—300.

every matrix with rational integral elements and of determinant 1 is a product of just three, namely

$$\begin{Vmatrix} 1 & 1 & \dots & 0 \\ 0 & 1 & \dots & 0 \\ \cdot & \cdot & \cdot & \cdot \\ 0 & 0 & \dots & 1 \end{Vmatrix}, \quad \begin{Vmatrix} 0 & -1 & \dots & 0 \\ 1 & 0 & \dots & 0 \\ \cdot & \cdot & \cdot & \cdot \\ 0 & 0 & \dots & 1 \end{Vmatrix}, \quad \begin{Vmatrix} 0 & 0 & \dots & 0 & (-1)^n \\ 1 & 0 & \dots & 0 & 0 \\ \cdot & \cdot & \cdot & \cdot & \cdot \\ 0 & 0 & \dots & 1 & 0 \end{Vmatrix}.$$

23. Greatest common divisors. If three matrices with elements in a principal ideal ring $\mathfrak{P}$ are in the relation $A = CD$, then D is called a *right divisor* of A, and A is called a *left multiple* of D. A *greatest common right divisor* (g.c.r.d.) D of two matrices A and B is a common right divisor which is a left multiple of every common right divisor of A and B. A *least common left multiple* (l.c.l.m.) of two matrices A and B is a common left multiple which is a right divisor of every common left multiple of A and B.

Theorem 23.1. Every pair of matrices A and B with elements in $\mathfrak{P}$ have a g.c.r.d. D expressible in the form

$$D = PA + QB.$$

Consider the matrix

$$F = \begin{Vmatrix} A & 0 \\ B & 0 \end{Vmatrix}$$

of order $2n$. As in the proof of Theorem 22.1, a unimodular matrix U of order $2n$ can be found such that the g.c.d. of the elements of the n-th column of F is in the (n, n)-position in UF. Then elementary transformations will reduce to 0 every element of this column below a_{nn}. This process may be continued to obtain an equation

$$\begin{Vmatrix} X_{11} & X_{12} \\ X_{21} & X_{22} \end{Vmatrix} \begin{Vmatrix} A & 0 \\ B & 0 \end{Vmatrix} = \begin{Vmatrix} D & 0 \\ 0 & 0 \end{Vmatrix}, \tag{23.1}$$

where the first factor X is unimodular. Thus

$$X_{11}A + X_{12}B = D$$

so that every common right divisor of A and B is a right divisor of D. Since X is unimodular, there exists a matrix $Y = X^{\mathrm{I}}$ with elements in $\mathfrak{P}$ such that

$$\begin{Vmatrix} A & 0 \\ B & 0 \end{Vmatrix} = \begin{Vmatrix} Y_{11} & Y_{12} \\ Y_{21} & Y_{22} \end{Vmatrix} \begin{Vmatrix} D & 0 \\ 0 & 0 \end{Vmatrix},$$

whence

$$A = Y_{11}D, \qquad B = Y_{21}D.$$

Hence D is a g.c.r.d. of A and B.

Corollary 23.11. If the $2n \cdot n$ array $\binom{A}{B}$ is of rank n, the matrices A and B have a non-singular g.c.r.d.

Corollary 23.12. If A and B have a non-singular g.c.r.d. D, every g.c.r.d. of A and B is of the form UD where U is unimodular.

For if D_1 is another g.c.r.d.,

$$D = PD_1, \quad D_1 = QD, \quad D = PQD, \quad PQ = I.$$

Theorem 23.2. Every pair of non-singular matrices A and B with elements in 𝔓 have a l.c.l.m. M unique up to a unit left factor.

The relation
$$X_{21}A + X_{22}B = O$$
follows from (23.1). Therefore
$$M = X_{21}A = -X_{22}B$$
is a common left multiple of A and B. If M_1 is another c.l.m. their g.c.r.d.
$$M_2 = PM + QM_1$$
is a c.l.m., so there exists a c.l.m. M_2 such that $M = HM_2$. Suppose
$$M_2 = KA = LB.$$
Then
$$X_{21}A = HKA, \quad -X_{22}B = HLB,$$
and since A and B are non-singular,
$$X_{21} = HK, \quad X_{22} = -HL.$$
But $I = X_{21}Y_{12} + X_{22}Y_{22} = H[KY_{12} - LY_{22}]$, so H is a unit matrix, and M is a right divisor of M_1.[1]

Lemma 23.3. Let A denote an n-th order matrix with elements in a proper euclidean ring 𝔈. For every element m of 𝔈 there exist matrices Q and R such that either (1) $A = mQ$ *or else* (2) $A = mQ + R$ *where*
$$0 < s[d(R)] < s(m^n).$$

By multiplying A on the left by a unimodular matrix, it can be put into Hermite's normal form. Determine $Q = (q_{rs})$, $R = (r_{rs})$ such that
$$a_{ij} = m q_{ij} + r_{ij},$$
where $0 \leqq s(r_{ij}) < s(m)$ for $i > j$, while $0 < s(r_{ij}) \leqq s(m)$ for $i = j$. Unless every
$$s(r_{ii}) = s(m),$$
$$s[d(R)] = s[r_{11} r_{22} \ldots r_{nn}] = s(r_{11}) s(r_{22}) \ldots s(r_{nn}) < [s(m)]^n = s(m^n),$$
and
$$A \overset{L}{=} mQ + R$$
is in form (2).

If every $s(r_{ii}) = s(m)$ and every other element of R is 0, $R = mE$ where E is diagonal, and
$$A \overset{L}{=} m(Q + E)$$
is in form (1).

If every $s(r_{ii}) = s(m)$ and some element of R below the main diagonal is not 0, e.g.,
$$R = \begin{Vmatrix} r_{11} & 0 & 0 \\ 0 & r_{22} & 0 \\ r_{31} & 0 & r_{33} \end{Vmatrix},$$

[1] The proofs of the last two theorems are due in essence to E. Cahen: Théorie des nombres Vol. I. Paris 1914, and in the form here presented to A. Châtelet: Groupes abéliens finis. Paris 1924.

it is always possible to obtain form (2). Let the last column having a non-zero element below the main diagonal be the i-th, and let r_{ki} be the first such element in that column. Add row k to row i. Then $UR = R_1 + mE_1$ where $s[d(R_1)] < s(m^n)$.

If $A \overset{L}{=} mQ_1 + R_1$, then $A = mUQ_1 + UR_1$ where $s[d(UR_1)] < s(m^n)$.

Theorem 23.3. If A and B are matrices with elements in a proper euclidean ring $\mathfrak{E}$, and if $d(B) \neq 0$, *there exist matrices Q and C such that $A = QB + C$ and either $C = 0$ or else $0 < s[d(C)] < s[d(B)]$.*

By the lemma there exist matrices Q and R such that

$$AB^{\mathrm{A}} = bQ + R, \qquad b = d(B),$$

and either $R = 0$ or else $0 < s[d(R)] < s(b^n)$. If $R = 0$, $A = QB$ and the theorem holds with $C = 0$. If $R \neq 0$, $R = AB^{\mathrm{A}} - bQ = AB^{\mathrm{A}} - QBB^{\mathrm{A}} = (A - QB)B^{\mathrm{A}} \equiv CB^{\mathrm{A}}$. Then

$$AB^{\mathrm{A}} = bQ + CB^{\mathrm{A}},$$
$$A = QB + C.$$

But $s[d(R)] = s[d(C)]\, s[d(B^{\mathrm{A}})] = s[d(C)]\, s[b^{n-1}] = s[d(C)]\, [s(b)]^{n-1}$. Therefore

$$0 < s[d(C)] < s(b) = s[d(B)].$$

The lemma and theorem were proved for rational integers by L. G. DU PASQUIER[1], who proceeded by this means to establish the existence of the g.c.r.d.

24. Linear form moduls. Let $\mathfrak{L}$ be a linear form modul of order n with respect to a ring $\mathfrak{R}$. That is, $\mathfrak{L}$ consists of all numbers of the form

$$a_1\varepsilon_1 + a_2\varepsilon_2 + \cdots a_n\varepsilon_n,$$

where the a's range independently over $\mathfrak{R}$, and the ε's are linearly independent with respect to $\mathfrak{R}$. The ε's are called a *basis* of $\mathfrak{L}$.

The basis is not unique, for any other set

$$\varepsilon_i' = \sum u_{ij}\varepsilon_j, \qquad (i, j = 1, 2, \ldots, n)$$

where $U = (u_{rs})$ is unimodular with elements in $\mathfrak{R}$ is also a basis for $\mathfrak{L}$; for every linear combination of the ε's is a linear combination of the ε''s, and vice versa. Conversely, every two bases of a linear form modul are related by such a transformation. The following discussion is relative to a fixed basis $\varepsilon_1, \varepsilon_2, \ldots, \varepsilon_n$ of $\mathfrak{L}$.

Let $\mathfrak{L}_1$ be a linear form submodul of order n of $\mathfrak{L}$,[2] and let $\mathfrak{L}_1$ have the basis $\lambda_1, \lambda_2, \ldots, \lambda_n$. Then

$$\lambda_i = \sum g_{ij}\varepsilon_j, \qquad (i, j = 1, 2, \ldots, n)$$

[1] DU PASQUIER, L. G.: Vjschr. naturforsch. Ges. Zürich Vol. 51 (1906) pp. 55 to 129.

[2] If $\mathfrak{R}$ is a principal ideal ring, every submodul of $\mathfrak{L}$ is a linear form modul. VAN DER WAERDEN: Moderne Algebra Vol. II p. 121. Berlin: Julius Springer 1931.

where $G_1 = (g_{rs})$ is a non-singular matrix with elements in $\mathfrak{R}$. We shall say that the matrix G_1 is *associated* with the basis $\lambda_1, \lambda_2, \ldots, \lambda_n$ of $\mathfrak{L}$. Every non-singular matrix G_1 determines in this way a basis of a linear form submodul of $\mathfrak{L}$.

If $\mathfrak{L}_2$ with basis $\mu_1, \mu_2, \ldots, \mu_n$ is a linear form submodul of order n of $\mathfrak{L}_1$, every number of $\mathfrak{L}_2$ is in $\mathfrak{L}_1$, and

$$\mu_i = \sum c_{ij}\lambda_j = \sum c_{ij} g_{jk} \varepsilon_k .$$

The matrix G_2 associated with the basis $\mu_1, \mu_2, \ldots, \mu_n$ of $\mathfrak{L}_2$ is CG_1, where C is a non-singular matrix with elements in $\mathfrak{R}$. This proves

Theorem 24.1. *$\mathfrak{L}_1$ contains $\mathfrak{L}_2$ if and only if G_1 is a right divisor of G_2.*

Corollary 24.1. *Two moduls $\mathfrak{L}_1$ and $\mathfrak{L}_2$ are equal if and only if $G_1 \stackrel{L}{=} G_2$.*

For if $G_2 = C_1G_1$ and $G_1 = C_2G_2$, then $C_1C_2 = I$ so that C_1 and C_2 are both unimodular.

We shall say that the matrix G_1 *corresponds* to the modul $\mathfrak{L}_1$ ($G_1 \backsim \mathfrak{L}_1$), understanding that G_1 is determined only up to a unit left factor.

Now specialize $\mathfrak{R}$ to a principal ideal ring $\mathfrak{P}$. The set of numbers common to two moduls $\mathfrak{L}_1$ and $\mathfrak{L}_2$ constitute a modul $\mathfrak{L}_d$ called the *greatest common submodul* (or logical product) of the two moduls $\mathfrak{L}_1$ and $\mathfrak{L}_2$. It may also be defined as that submodul of $\mathfrak{L}_1$ and $\mathfrak{L}_2$ which contains every common submodul of $\mathfrak{L}_1$ and $\mathfrak{L}_2$.

The set of all numbers contained in either $\mathfrak{L}_1$ or $\mathfrak{L}_2$ or both, together with their sums and differences, constitutes a modul $\mathfrak{L}_m$ called the *least common supermodul* (or logical sum) of $\mathfrak{L}_1$ and $\mathfrak{L}_2$. It may also be defined as that modul containing $\mathfrak{L}_1$ and $\mathfrak{L}_2$ which is contained in every modul containing $\mathfrak{L}_1$ and $\mathfrak{L}_2$.

Theorem 24.2. *Let $G_1 \backsim \mathfrak{L}_1$, $G_2 \backsim \mathfrak{L}_2$, $G_d \backsim \mathfrak{L}_d$, $G_m \backsim \mathfrak{L}_m$. Then G_d is the g.c.r.d. of G_1 and G_2, and G_m is the l.c.l.m.* of *G_1 and G_2.*

It is proper to speak of *the* g.c.r.d. and l.c.l.m. for each is determined up to a unit left factor, the same latitude of definition as that of G_d and G_m. The proof follows directly from Theorem 24.1.

The application of matrices with rational integral elements to the theory of moduls is in large part due to *A.* CHÂTELET[1]. His work is summarized in two books, Leçons sur la théorie des nombres, Paris 1913, and Groupes abéliens finis, Paris 1924.

25. Ideals. Let $\mathfrak{R}$ be a ring with unit element, and let $\mathfrak{S}$ be a linear form modul with respect to $\mathfrak{R}$ which is also a ring, and whose elements are commutative with those of $\mathfrak{R}$. An instance of such a system is a domain of integrity of a linear associative algebra in the sense of DICKSON[2].

[1] CHÂTELET, A.: e.g., Ann. École norm. III Vol. 28 (1911) pp. 105—202 — C. R. Acad. Sci., Paris Vol. 154 (1912) p. 502.

[2] DICKSON, L. E.: Algebren und ihre Zahlentheorie, p. 154. Zürich 1927.

As in § 1 let the constants of multiplication of $\mathfrak{S}$ be c_{ijk}, and define

$$R_i = (c_{isr}).$$

A submodul of $\mathfrak{S}$ which is closed under multiplication on the left by numbers of $\mathfrak{S}$ is called a *left ideal*[1]. Similarly there may be defined *right ideals* and *two-sided ideals.*

Theorem 25.1. A modul $\mathfrak{L}$ of $\mathfrak{S}$ is a left ideal if and only if its corresponding matrix G satisfies the conditions

$$G R_i^{\mathrm{T}} = D_i G, \qquad (i = 1, 2, \ldots, n)$$

where the D_i are matrices with elements in $\mathfrak{R}$.

Assume that $\lambda_1, \lambda_2, \ldots, \lambda_n$ constitute a basis for a left ideal $\mathfrak{J}$, where

$$\lambda_i = \sum g_{ij} \varepsilon_j.$$

Every number k of $\mathfrak{J}$ is of the form

$$k = \sum k_i \lambda_i = \sum k_i g_{ij} \varepsilon_j,$$

while every number s of $\mathfrak{S}$ is of the form

$$s = \sum s_l \varepsilon_l.$$

Since sk is in $\mathfrak{J}$ for every s_l and k_i, there exist numbers d_r of $\mathfrak{R}$ such that

$$sk = \sum s_l k_i g_{ij} c_{ljh} \varepsilon_h = \sum d_r g_{rt} \varepsilon_t.$$

Since the ε's are linearly independent,

$$\sum s_l k_i g_{ij} c_{ljh} = \sum d_r g_{rh}.$$

In particular there exist values d_{pqr} of d_r when $s_l = \delta_{lp}$ and $k_i = \delta_{iq}$. For these values

$$\sum g_{qj} c_{pjh} = \sum d_{pqr} g_{rh},$$

or

$$G R_p^{\mathrm{T}} = D_p G, \quad D_p = (d_{prs}). \qquad (p = 1, 2, \ldots, n)$$

Conversely, the condition is sufficient. Let d_{pqr} and g_{qj} be numbers of $\mathfrak{R}$ satisfying the above conditions. Define

$$\lambda_i = \sum g_{ij} \varepsilon_j.$$

The set of numbers $k = \sum k_i \lambda_i$ is evidently a modul. Let s be any number of $\mathfrak{S}$. Then

$$sk = \sum s_l k_i g_{ij} c_{ljr} \varepsilon_r = \sum s_l k_i d_{lis} \lambda_s$$

is again in the modul, which is therefore a left ideal[2].

CHÂTELET[3] set up a correspondence between ideals in an algebraic field and matrices with rational integral elements. If $\lambda_1, \lambda_2, \ldots, \lambda_n$

[1] VAN DER WAERDEN: Moderne Algebra Vol. I p. 53.
[2] MACDUFFEE, C. C.: Trans. Amer. Math. Soc. Vol. 31 (1929) pp. 71–90.
[3] CHÂTELET: Ann. École norm. III Vol. 28 (1911) pp. 105–202.

are the conjugates of λ in $\mathfrak{F}(\lambda)$, and $E = [\lambda_1, \lambda_2, \ldots, \lambda_n]$ is a diagonal matrix, he proved that a necessary and sufficient condition that AEA^{I} be integral is that A, apart from a diagonal matrix as a factor, correspond to a basis of an ideal.

Now specialize $\mathfrak{R}$ to a principal ideal ring $\mathfrak{P}$. If two left ideals $\mathfrak{J}_1$ and $\mathfrak{J}_2$ of $\mathfrak{S}$ have bases $\lambda_1, \ldots, \lambda_n$ and $\mu_1, \ldots, \mu_n$ respectively, the set of numbers

$$\sum d_{ij} \lambda_i \mu_j ,$$

where d_{ij} range over $\mathfrak{P}$ is evidently a modul. Since $\mathfrak{P}$ is a principal ideal ring, it is a linear form modul. It is readily seen to be closed under multiplication on the left by a number of $\mathfrak{S}$, so it is a left ideal. The ideal so defined is called the product of the ideals $\mathfrak{J}_1$ and $\mathfrak{J}_2$ in that order.

H. POINCARÉ[1] set up a correspondence between matrices and quadratic ideals which is an instance of the correspondence of this paragraph, and called the matrix corresponding to the ideal product the *second product* or *commutative product* of the matrices corresponding to the factors. He noted the isomorphism of ideal multiplication with composition of quadratic forms.

The matrix corresponding to the product of the two left ideals $\mathfrak{J}_1$ and $\mathfrak{J}_2$ is the g.c.r.d. of the matrices

$$G_2 R^{\mathrm{T}}(\lambda_i), \qquad (i = 1, 2, \ldots, n)$$

where G_2 corresponds to $\mathfrak{J}_2$, and $\lambda_1, \lambda_2, \ldots, \lambda_n$ is a basis of I_1.[2]

If $\alpha_1, \ldots, \alpha_k$ are numbers of an algebraic field of order n, a minimal basis of the ideal $(\alpha_1, \ldots, \alpha_k)$ is readily determined from its associated matrix A, which is a g.c.r.d. of

$$R(\alpha_1),\ R(\alpha_2),\ \ldots,\ R(\alpha_k).$$[3]

IV. Equivalence.

26. Equivalent matrices. Let $A = PBQ$, where each matrix has its elements in a principal ideal ring $\mathfrak{P}$. Then A is a *multiple* of B.[4]

Theorem 26.1. If A is a multiple of B, the g.c.d. d_i of the i-rowed minor determinants of B divides the g.c.d. d_i' of the i-rowed minor determinants of A.

This follows immediately from Theorem 7.9.

Two matrices A and B with elements in $\mathfrak{P}$ are *equivalent* $(A \overset{\mathrm{E}}{=} B)$ if there exist two unimodular matrices U and V such that $A = UBV$.

---

[1] POINCARÉ, H.: J. École polytechn. Cah. 47 (1880) pp. 177–245.

[2] SHOVER, GRACE, and C. C. MACDUFFEE: Bull. Amer. Math. Soc. Vol. 37 (1931) pp. 434–438.

[3] MACDUFFEE, C. C.: Math. Ann. Vol. 105 (1931) pp. 663–665.

[4] HENSEL, K.: J. reine angew. Math. Vol. 114 (1894) pp. 109–115.

The relation of equivalence is determinative, reflexive, symmetric and transitive. (Cf. § 22.) The present chapter has to do with those properties of matrices which are invariant under this relationship.

Corollary 26.1. If $A \overset{E}{=} B$, *every g.c.d.* d_i *of the i-rowed minor determinants of B is associated with every g.c.d.* d_i' *of the i-rowed minor determinants of A.*[1]

Elementary operations on the rows of a matrix are defined as in § 22, each being accomplished by multiplying the matrix on the left by a unimodular matrix. Elementary operations on the columns are defined in an analogous manner, each being accomplished by multiplying the matrix on the right by a unimodular matrix. The inverse of an elementary operation is an elementary operation of the same type.

Theorem 26.2 Every matrix A of rank ϱ *with elements in* $\mathfrak{P}$ *is equivalent to a diagonal matrix* $[h_1, h_2, \ldots, h_\varrho, 0, \ldots, 0]$ *where* $h_i | h_{i+1}$.[2]

This diagonal form will be called SMITH's *normal form.*

If A is of rank ϱ, the rows and columns can be shifted by elementary transformations so that the minor determinant of order ϱ in the upper left corner is $\neq 0$. Then as in the proof of Theorem 22.1, the element in the (1, 1)-position can be made $\neq 0$ and a g.c.d. of the elements of the first column. The elements of the first column below the first row can then be made 0's by elementary transformations on the rows. If the element which now stands in the (1, 1)-position divides every other element of the first row, these other elements can all be made 0's by elementary transformations on the columns so as not to disturb the first column of 0's. If they are not all divisible by this element a_{11}, then a_{11} can be replaced by the g.c.d. of the elements of the first row, and this g.c.d. will contain fewer prime factors than a_{11}. The process is now repeated until an element in the (1, 1)-position is obtained which divides every other element of the first row and every other element of the first column. Since every number of $\mathfrak{P}$ is factorable into a finite number of primes, this stage is reached in a finite number of steps[3].

By working with the last $n-1$ rows and columns, then with the last $n-2$ rows and columns etc., A can be reduced to an equivalent matrix

$$\left\| \begin{matrix} D & 0 \\ 0 & M \end{matrix} \right\|, \qquad D = [h_1, h_2, \ldots, h_\varrho], \qquad h_i \neq 0.$$

Now $M = 0$, for if one element of M were not 0, it could be shifted into the $(\varrho+1, \varrho+1)$-position, and A would have a non-vanishing minor determinant of order $\varrho + 1$.

---

[1] SYLVESTER, J. J.: Philos. Mag. Vol. 1 (1851) pp. 119–140.

[2] SMITH, H. J. S.: Philos. Trans. Roy. Soc. London Vol. 151 (1861–1862) p. 314.

[3] VAN DER WAERDEN: Moderne Algebra Vol. II p. 124. Berlin: Julius Springer 1931.

By adding column 2, column 3, ..., column ϱ to column 1, D is made to assume the form

$$\left\| \begin{matrix} h_1 & 0 & 0 & \dots & 0 \\ h_2 & h_2 & 0 & \dots & 0 \\ h_3 & 0 & h_3 & \dots & 0 \\ \cdot & \cdot & \cdot & \cdot & \cdot \\ h_\varrho & 0 & 0 & \dots & h_\varrho \end{matrix} \right\| .$$

As in the proof of Theorem 22.1, there is a unimodular matrix U which, used as a left factor, replaces h_1 by the g.c.d. of $h_1, \dots, h_\varrho$. The new matrix UD has every element a homogeneous linear combination of $h_1, \dots, h_\varrho$, so each element of UD is divisible by the new h_1. Again reduce to the diagonal form $[h_1, h_2, \dots, h_\varrho]$ where now h_1 divides $h_2, \dots, h_\varrho$. Continue until $h_i \mid h_{i+1}$, $i = 1, 2, \dots, \varrho - 1$.

Now consider a matrix A with elements in a ring $\mathfrak{E}'$, with unit element and no divisors of zero, in which both left and right division transformations exist. That is, a stathm is defined for every number of $\mathfrak{E}'$ except 0, and for every pair of numbers a and b, $b \neq 0$, there exist numbers q, q', r, r', such that

$$a = bq + r, \qquad r = 0 \quad \text{or} \quad s(r) < s(b),$$
$$a = q'b + r', \qquad r' = 0 \quad \text{or} \quad s(r') < s(b).$$

(Cf. § 20.) If $\mathfrak{E}'$ is commutative, it is a euclidean principal ideal ring $\mathfrak{E}$. If $s(ab) = s(a)\,s(b)$, $\mathfrak{E}'$ is proper.

An elementary transformation is one of five types:

1. The addition to the elements of any row of the products of any element k of $\mathfrak{E}'$ by the corresponding elements of another row, k being used as a left factor.

2. The addition to the elements of any column of the products of any element k of $\mathfrak{E}'$ by the corresponding elements of another column, k being used as a right factor.

3. The interchange of two rows or of two colums.

4. The insertion of the same unit factor before each element of any row.

5. The insertion of the same unit factor after each element of any column.

Each of these elementary operations can be effected by multiplying the given matrix either on the left or on the right by a certain elementary matrix. A product of elementary matrices is called a *unit matrix*, in spite of the fact that the concept of determinant is not defined for matrices with elements in a non-commutative ring.

J. H. M. Wedderburn[1] has proved that if A has elements in $\mathfrak{E}'$, there exist unit matrices P and Q such that

$$PAQ = [h_1, h_2, \dots, h_\varrho, 0, \dots, 0],$$

[1] Wedderburn, J. H. M.: J. reine angew. Math. Vol. 167 (1931) pp. 129—141.

ϱ being defined as the rank of A, and if $\mathfrak{E}'$ is proper, h_i is both a right and a left divisor of h_{i+1}.[1]

This last result for $\mathfrak{E}'$ proper had been given essentially by L. E. DICKSON[2]. Since the resulting diagonal matrices are factorable into prime matrices in but one way apart from unit factors, the same is true for all matrices with elements in $\mathfrak{E}'$ which are of rank n.

27. Invariant factors and elementary divisors. Let A be a matrix with elements in a principal ideal ring $\mathfrak{P}$, and let

$$D = [h_1, h_2, \ldots, h_\varrho, 0, \ldots, 0]$$

be its equivalent normal form (Theorem 26.2). It has been seen (Theorem 26.1) that the g.c.d. d_i of the i-rowed minor determinants of A is associated with the g.c.d. d_i' of the i-rowed minor determinants of D. Since $h_i | h_{i+1}$, it follows that

Theorem 27.1. The g.c.d. $d_i = h_1 h_2 \ldots h_i$ *of the i-rowed minor determinants of* A *divides the g.c.d.* $d_{i+1} = h_1 h_2 \ldots h_{i+1}$ *of the* $(i+1)$-*rowed minor determinants of* A.

The quotients $h_1 = d_1$, $h_2 = d_2/d_1$, $h_3 = d_3/d_2, \cdots$ are called the *invariant factors* of A. They are invariants under the relation of equivalence, and are determined up to a unit factor.

In a principal ideal ring every element neither 0 nor a unit can be factored uniquely (except for unit factors) into a product of powers of primes[3]. Suppose

$$h_i = p_1^{l_{i1}}, p_2^{l_{i2}}, \ldots, p_k^{l_{ik}}.$$

Since $h_i | h_{i+1}$, the exponents of each prime factor form a sequence

$$e_{nl} \geqq e_{n-1,l} \geqq \cdots \geqq e_{1l}. \qquad (l = 1, 2, \ldots, k)$$

Such of these powers $p_l^{l_{il}}$ as are not units are called the *elementary divisors* of A [WEIERSTRASS]. They are defined up to a unit factor.

The elementary divisors are *simple* if each is a prime[4].

Theorem 27.2. $A \overset{E}{=} B$ *if and only if* A *and* B *have the same elementary divisors (invariant factors).*

If A and B have the same invariant factors, they can be reduced to the same normal form and hence are equivalent. If they are equivalent, they have the same invariant factors, for these are invariants. The invariant factors determine the elementary divisors uniquely, and conversely.

Corollary 27.2. Two matrices A *and* B *with elements in a commutative field* $\mathfrak{F}$ *are equivalent if and only if they have the same rank.*

[1] See also VAN DER WAERDEN: l. c.

[2] DICKSON, L. E.: Algebras and their Arithmetics. Univ. of Chicago Press 1923 p. 174.

[3] VAN DER WAERDEN: Moderne Algebra Vol. I p. 65.

[4] KRONECKER, L.: M.-B. preuß. Akad. Wiss. 1874.

For in this case every element of $\mathfrak{F}$ except 0 is a unit, and a normal form $[1, \ldots, 1, 0, \ldots, 0]$ may be chosen where the number of 1's is the rank ϱ.

The theory of elementary divisors is one of the oldest and most thoroughly exploited branches of matric theory. The literature is so extensive that the reader is referred to P. MUTH's *Theorie und Anwendungen der Elementarteiler*[1] for the early papers. The theory was initiated by K. WEIERSTRASS[2] for the polynomial domain of the complex field, by H. J. S. SMITH[3] and G. FROBENIUS[4] for matrices with rational integral elements, and by FROBENIUS[5] for matrices with elements in a modular field. FROBENIUS[6] later gave a rational treatment of the WEIERSTRASS theory.

A. CHÂTELET[7] proved that if A and B have rational integral elements, and if D is their g.c.r.d. and M their l.c.l.m., then MA^{I} and MB^{I} have, respectively, the same invariant factors as BD^{I} and AD^{I}.

28. Factorization of a matrix. Let A be a matrix of rank ϱ with elements in a principal ideal ring $\mathfrak{P}$. By Theorem 26.2, A is a product by unit matrices of a diagonal matrix

$$D = [h_1, h_2, \ldots, h_\varrho, 0, \ldots, 0].$$

By § 4, D is a product of matrices of the type

$$[1, 1, \ldots, p_i, \ldots, 1],$$

where p_i is an irreducible factor of h_i, and a diagonal matrix $[1, 1, \ldots, 1, 0, \ldots, 0]$ of rank ϱ. The matrices $[1, \ldots, p_i, \ldots, 1]$ have prime determinants and are therefore irreducible[8].

L. KRONECKER[9] gave the following decomposition for a unimodular matrix with elements in a field for the case $\alpha \neq 0$:

$$\begin{pmatrix} 1 & -\gamma/\alpha \\ 0 & 1 \end{pmatrix} \begin{pmatrix} 0 & -1 \\ 1 & 0 \end{pmatrix}^3 \begin{pmatrix} \alpha & 0 \\ 0 & 1/\alpha \end{pmatrix} \begin{pmatrix} 0 & -1 \\ 1 & 0 \end{pmatrix} \begin{pmatrix} 1 & \beta/\alpha \\ 0 & 1 \end{pmatrix} = \begin{pmatrix} \alpha & \beta \\ \gamma & \dfrac{\beta\gamma+1}{\alpha} \end{pmatrix}$$

and another decomposition when $\alpha = 0$.

H. LAURENT[10] discussed the factorization of a matrix into elementary matrices.

[1] MUTH, P.: Teubner 1899.
[2] WEIERSTRASS, K.: M.-B. preuß. Akad. Wiss. 1868 pp. 310–338.
[3] SMITH, H. J. S.: Proc. London Math. Soc. Vol. 4 (1873) pp. 236–253.
[4] FROBENIUS, G.: J. reine angew. Math. Vol. 86 (1879) pp. 146–208.
[5] FROBENIUS, G.: J. reine angew. Math. Vol. 88 (1880) pp. 96–116.
[6] FROBENIUS, G.: S.-B. preuß. Akad. Wiss. 1894 pp. 31–44.
[7] CHÂTELET, A.: C. R. Acad. Sci., Paris Vol. 177 (1923) pp. 729–731.
[8] DU PASQUIER, L. G.: Vjschr. naturforsch. Ges. Zürich Vol. 51 (1906) pp. 55 to 129.
[9] KRONECKER, L.: S.-B. preuß. Akad. Wiss. 1889 pp. 479–505.
[10] LAURENT, H.: Nouv. Ann. Math. III Vol. 15 (1896) pp. 345–365.

J. WELLSTEIN[1] showed that every matrix is a product of elementary matrices of three types: $E_{ij}(x)$ is obtained by replacing the 0 in row i and column j of the identity matrix by x. $E_k(x)$ is obtained by replacing the 1 in row k and column k of the identity matrix by x. Γ is the identity matrix with the rows cyclicly permuted.

C. CELLITTI[2] noted that every second order integral matrix is a product of powers of

$$\begin{pmatrix} 1 & 1 \\ 0 & 1 \end{pmatrix}, \quad \begin{pmatrix} 1 & 0 \\ 1 & 1 \end{pmatrix}, \quad \begin{pmatrix} a & 0 \\ 0 & 1 \end{pmatrix}.$$

29. Polynomial domains. An important instance of a principal ideal ring is the polynomial domain $\mathfrak{P}(\lambda)$ of all polynomials in λ with coefficients in a commutative field $\mathfrak{F}$. The elements of $\mathfrak{F}$ (except 0) are the units of $\mathfrak{P}(\lambda)$. Moreover $\mathfrak{P}(\lambda)$ is euclidean, for if a and $b \neq 0$ are two numbers of the domain, there exist two other numbers q and r such that

$$a = bq + r,$$

where either $r = 0$ or else r is of lower degree in λ than q. If $\mathfrak{F}$ is algebraically closed[3] (e.g., the complex field), the primes in $\mathfrak{P}(\lambda)$ are the linear polynomials in λ.

A matrix $A = (a_{rs0} + a_{rs1}\lambda + \cdots + a_{rsk}\lambda^k)$ with elements in $\mathfrak{P}(\lambda)$ can be written as a polynomial in λ,

$$A = (a_{rs0}) + (a_{rs1})\lambda + \cdots + (a_{rsk})\lambda^k,$$

whose coefficients are matrices with elements in $\mathfrak{F}$. The matrix A is of *degree* k if $(a_{rsk}) \neq 0$. It is *proper* if it is of degree k and $d(a_{rsk}) \neq 0$.

Theorem 29.1. If A and B are matrices with elements in $\mathfrak{P}(\lambda)$, and if B is proper of degree l, then there exist matrices Q and R (Q_1 and R_1) such that

$$A = BQ + R, \qquad A = Q_1 B + R_1,$$

where either $R = 0$ ($R_1 = 0$) or else $R(R_1)$ is of degree $< l$.

Let

$$A = A_k\lambda^k + \cdots + A_0, \qquad B = B_l\lambda^l + \cdots + B_0, \qquad l < k.$$

Since $d(B_l) \neq 0$, the equation $B_l X = A_k$ has a solution $X = C_{k-l}$. Then $A - BC_{k-l}\lambda^{k-l}$ is of degree $k-1$ at most. Continue as in ordinary long division until a remainder is obtained of degree $< l$.

Theorem 29.2. If A and B are proper of degrees k and l respectively, and if $AP_1 = P_2 B$, there exists a matrix Q and two matrices R_1 and R_2, of degrees r_1 and r_2 respectively, such that

$$AR_1 = R_2 B, \quad P_1 = QB + R_1, \quad P_2 = AQ + R_2, \quad r_1 < l, \quad r_2 < k.$$[4]

[1] WELLSTEIN, J.: Nachr. Ges. Wiss. Göttingen 1909 pp. 77–99.

[2] CELLITTI, C.: Atti Accad. naz. Lincei, Rend. V Vol. 23 II (1914) pp. 208–212.

[3] VAN DER WAERDEN: Moderne Algebra Vol. I p. 198.

[4] FROBENIUS, G.: S.-B. preuß. Akad. Wiss. 1910 pp. 3–15.

Determine Q_1, Q_2, R_1, R_2 so that

$$P_2 = AQ_2 + R_2, \qquad P_1 = Q_1 B + R_1,$$

where R_1 is O or of degree $< l$, and R_2 is O or of degree $< k$. Then

$$AR_1 - R_2 B = A(Q_2 - Q_1)B.$$

The left member is either O or of degree $< k+l$ while the right member is either O or of degree $\geqq k+l$. Hence $AR_1 = R_2 B$ and, since A and B are non-singular, $Q_2 = Q_1$.

Theorem 29.3. If A and B are proper, of degree 1, and equivalent, then there exist non-singular matrices P and Q with elements in $\mathfrak{F}$ such that $A = PBQ$.

Since A and B are equivalent in $\mathfrak{P}(\lambda)$, there exist non-singular matrices P_1 and P_2 whose determinants are in $\mathfrak{F}$ such that

$$AP_1 = P_2 B.$$

Since A and B are each of degree 1, the matrices R_1 and R_2 of Theorem 29.2 have elements in $\mathfrak{F}$. It remains only to show that R_1 and R_2 are non-singular. Let

$$P_1^{\mathrm{I}} = Q_3 A + R_3,$$

where R_3 is either O or of degree $r_3 = 0$. Then

$$\begin{aligned} I = P_1^{\mathrm{I}} P_1 &= (Q_3 A + R_3)(QB + R_1) \\ &= Q_3 AQB + Q_3 AR_1 + R_3 QB + R_3 R_1, \\ I - R_3 R_1 &= (Q_3 AQ + Q_3 R_2 + R_3 Q) B. \end{aligned}$$

The left member is either O or of degree 0 in λ, while the right member is O or of degree $\geqq 1$ in λ. Hence each member is O, $R_2 R_1 = I$, $d(R_1) \neq 0$. Similarly $d(R_2) \neq 0$. Then $A = PBQ$, where $P = R_2$ and $Q = R_1^{\mathrm{I}}$.

J. A. DE SEGUIER[1] showed how to reduce $A_1\lambda + A_2\mu$ to diagonal form directly by means of constant matrices P and Q.

The characteristic matrix $I\lambda - A$ of the matrix A with elements in a commutative field $\mathfrak{F}$ is an important instance of a matrix in $\mathfrak{P}(\lambda)$ whose elements are linear. Let

$$I\lambda - A \overset{\mathrm{E}}{=} [h_1(\lambda), h_2(\lambda), \ldots, h_n(\lambda)],$$

where the second member is the normal form of Theorem 26.2.

Theorem 29.4. The minimum equation of A is $h_n(\lambda) = 0$.

This follows from Theorems 15.1 and 27.1.

Corollary 29.4. A matrix is not derogatory when and only when the elementary divisors of its characteristic matrix are powers of distinct primes.

[1] DE SEGUIER, J. A.: Bull. Soc. Math. France Vol. 36 (1908) pp. 20—40.

A matrix of the form

$$A = \begin{Vmatrix} A_1 & 0 & \dots & 0 \\ 0 & A_2 & \dots & 0 \\ \cdot & \cdot & \cdot & \cdot \\ 0 & 0 & \dots & A_k \end{Vmatrix}$$

will be called the *direct sum* of the matrices $A_1, A_2, \dots, A_k$, and will be written[1]

$$A = A_1 \dotplus A_2 \dotplus \cdots \dotplus A_k .$$

Theorem 29.5. $I\lambda - A$ *is equivalent to the matrix*

$$B(\lambda) = B_n(\lambda) \dotplus B_{n-1}(\lambda) \dotplus \cdots \dotplus B_{n-k}(\lambda),$$

where $[1, \dots, 1, h_{n-k}(\lambda), \dots, h_n(\lambda)]$ *is the* SMITH *normal form of* $I\lambda - A$, *and* $B_i(\lambda)$ *is any matrix whose invariant factors are all* 1*'s but the last which is* $h_i(\lambda)$.

Suppose $P_i B_i(\lambda) Q_i = [1, \dots, 1, h_i(\lambda)]$. If

$$P = P_n \dotplus P_{n-1} \dotplus \cdots \dotplus P_{n-k}, \qquad Q = Q_n \dotplus Q_{n-1} \dotplus \cdots \dotplus Q_{n-k},$$

then a suitable permutation of the rows and columns of PBQ gives $[1, \dots, 1, h_{n-k}, \dots, h_n]$.

In particular $B_i(\lambda)$ may be chosen to be

$$\begin{Vmatrix} \lambda & -1 & 0 & \dots & 0 \\ 0 & \lambda & -1 & \dots & 0 \\ \cdot & \cdot & \cdot & \cdot & \cdot \\ h_l & h_{l-1} & h_{l-2} & \dots & h_1 \end{Vmatrix},$$

where $h_i(\lambda) = \lambda^l + h_1\lambda^{l-1} + \cdots + h_l$.

If $h_i(\lambda)$ is completely reducible in $\mathfrak{F}$, say

$$h_i(\lambda) = (\lambda - \lambda_1)^{l_{i1}} \dots (\lambda - \lambda_k)^{l_{ik}},$$

$B_i(\lambda)$ may be chosen in the form $I\lambda - J$, where J is the JORDAN form discussed in § 14. (See Corollary 15.1.)

If $\mathfrak{F}$ is an algebraically closed field $\mathfrak{C}$, every $h_i(\lambda)$ being completely reducible, the JORDAN normal form corresponding to each invariant factor is a direct sum of JORDAN forms each corresponding to an elementary divisor of $I\lambda - A$. By a shifting of the rows and columns these forms can be arranged in any order. Thus $B(\lambda)$ can be chosen

$$B(\lambda) = J_{11}(\lambda) \dotplus J_{12}(\lambda) \dotplus \cdots \dotplus J_{nk}(\lambda),$$

where each $J_{il}(\lambda)$ is the JORDAN form corresponding to an elementary divisor $(\lambda - \lambda_l)^{l_{il}}$.

It will be seen that there are two types of invariant for the field $\mathfrak{C}$, the distinct roots of the characteristic equation, $\lambda_1, \dots, \lambda_j$, which may

[1] KREIS, H.: Contribution à la théorie des systèms linéaires. Zürich 1906. KREIS attributes the notation to A. HURWITZ.

be called *numerical invariants*, and the exponents e_{il} of the elementary divisors, which may be called *invariants of structure*[1]. The exponents written in the array

$$\begin{matrix} e_{11} & e_{21} & \cdots & e_{j1} \\ e_{12} & e_{22} & \cdots & e_{j2} \\ \cdot & \cdot & \cdot & \cdot \end{matrix}$$

or $[(e_{11}, e_{12}, \ldots), (e_{21}, e_{22}, \ldots), \ldots]$ constitute the SEGRE *characteristic* of the matrix[2].

30. Equivalent pairs of matrices. Two pairs of matrices A_1, A_2 and B_1, B_2 with elements in a commutative field $\mathfrak{F}$ are said to be *equivalent* if and only if there exist two non-singular matrices P and Q with elements in $\mathfrak{F}$ such that

$$A_1 = PB_1Q, \quad A_2 = PB_2Q.$$

Theorem 30.1. If A_1 and B_1 are non-singular, the pairs of matrices A_1, A_2 and B_1, B_2 are equivalent if and only if the matrices $A_1\lambda + A_2$ and $B_1\lambda + B_2$ in the polynomial domain $\mathfrak{P}(\lambda)$ have the same invariant factors (or elementary divisors)[3].

Let $A_1\lambda + A_2 = A$, $B_1\lambda + B_2 = B$. If $A_1 = PB_1Q$ and $A_2 = PB_2Q$, then evidently $A = PBQ$ for every λ. In the domain $\mathfrak{P}(\lambda)$, P and Q are unimodular since their determinants are non-zero numbers of $\mathfrak{F}$, so by Theorem 27.2 the invariant factors of A and B coincide.

If, conversely, A and B have the same invariant factors, there exist two matrices P_1 and Q_1 whose determinants are non-zero numbers of $\mathfrak{F}$, but whose elements may involve λ, such that $A = P_1BQ_1$. Since A and B are proper, there exist by Theorem 29.3 two non-singular matrices P and Q with elements in $\mathfrak{F}$ such that

$$A_1\lambda + A_2 = P(B_1\lambda + B_2)Q$$

for λ indeterminate. Hence $A_1 = PB_1Q$ and $A_2 = PB_2Q$.

In treating the case where both A_1 and A_2 are singular, it is more convenient to use the symmetric linear combination $A_1\lambda + A_2\mu$ whose elements lie in the polynomial domain $\mathfrak{P}(\lambda, \mu)$ of homogeneous polynomials with coefficients in $\mathfrak{F}$. $\mathfrak{P}(\lambda, \mu)$ is isomorphic with $\mathfrak{P}(\lambda)$. The invariant factors of $A_1\lambda + A_2\mu$ will be called the invariant factors of the pair A_1, A_2. The pair of matrices A_1, A_2 is said to be a *non-singular pair* if $d(A_1\lambda + A_2\mu)$ is not zero in $\mathfrak{P}(\lambda, \mu)$.

Lemma 30.2. If $A_1' = A_1p + A_2q$, $A_2' = A_1r + A_2s$, where p, q, r, s and the elements of A_1 and A_2 are in $\mathfrak{F}$, and if $ps - rq \neq 0$, then the invariant factors of $A_1'\lambda + A_2'\mu$ are obtained from those of $A_1u + A_2v$ by the substitution

$$u = p\lambda + r\mu, \quad v = q\lambda + s\mu.$$

[1] Cf. S. LATTÈS: Ann. Fac. Sci. Univ. Toulouse Vol. 28 (1914) pp. 1–84.

[2] SEGRE, C.: Atti Accad. naz. Lincei, Mem. III Vol. 19 (1884) pp. 127–148.

[3] WEIERSTRASS: M.-B. preuß. Akad. Wiss. 1868 pp. 310–338.

This substitution, having an inverse, defines an automorphism of the domain $\mathfrak{P}(\lambda, \mu) \sim \mathfrak{P}(u, v)$ by which $A_1'\lambda + A_2'\mu \sim A_1 u + A_2 v$.

Theorem 30.2. Two non-singular pairs of matrices A_1, A_2 and B_1, B_2 with elements in a field $\mathfrak{F}$ are equivalent if and only if they have the same invariant factors.

If they are equivalent, they have the same invariant factors. This follows as in the proof of Theorem 30.1.

If A_1, A_2 is a non-singular pair, there exists a non-singular matrix $A_1' = A_1 p + A_2 q$. Choose r and s in any way so that $ps - rq \neq 0$, and define $A_2' = A_1 r + A_2 s$. Define B_1', B_2' cogrediently. Then by the lemma A_1', A_2' have the same invariant factors as B_1', B_2', since A_1, A_2 have the same invariant factors as B_1, B_2. In this case the pairs A_1', A_2' and B_1', B_2' are equivalent by Theorem 30.1. Then the pairs A_1, A_2 and B_1, B_2 are equivalent.

Corollary 30.21. If A_1 is non-singular, the pair A_1, A_2 with elements in $\mathfrak{F}$ is equivalent to the canonical pair $I, -B$ where

$$B = B_n \dot{+} B_{n-1} \dot{+} \cdots \dot{+} B_{n-k},$$

and B_i is the companion matrix of the i-th invariant factor of A_1, A_2.

Corollary 30.22. If A_1 is non-singular the pair A_1, A_2 with elements in an algebraically closed field $\mathfrak{C}$ is equivalent to the canonical pair $I, -J$ where

$$J = J_n \dot{+} J_{n-1} \dot{+} \cdots \dot{+} J_{n-k},$$

and J_i is the JORDAN *matrix of the i-th invariant factor of A_1, A_2.*

The problem of the equivalence of singular pairs of matrices presents more difficulty. M. PASCH[1] and P. MUTH[2] treated singular pairs of third order matrices. MUTH[3] treated the general case. L. E. DICKSON[4] proved that two singular pairs are equivalent by rational transformations if and only if they have the same invariant factors and the same *minimal numbers* M_i, which he defines. TURNBULL and AITKEN[5] have an original treatment of the singular case.

Since a polynomial in more than one variable is usually not factorable into linear factors, the WEIERSTRASS elementary divisor theory does not generalize so as to be applicable to the problem of the equivalence of sets of more than two matrices. S. KANTOR[6] generalized the concept of elementary divisor by geometric methods to handle this problem.

A new method in the equivalence of pairs of matrices was developed by R. G. D. RICHARDSON[7]. First suppose that $\lambda_1, \ldots, \lambda_n$ are the roots

[1] PASCH, M.: Math. Ann. Vol. 38 (1891) pp. 24–49.
[2] MUTH, P.: Math. Ann. Vol. 42 (1893) pp. 257–272.
[3] MUTH, P.: Theorie und Anwendungen der Elementarteiler. Teubner 1899.
[4] DICKSON, L. E.: Trans. Amer. Math. Soc. Vol. 29 (1927) pp. 239–253.
[5] TURNBULL and AITKEN: Canonical matrices, Chap. IX. Glasgow 1932.
[6] KANTOR, S.: S.-B. Bayer. Akad. Wiss. Vol. 98 (1897) pp. 367–381.
[7] RICHARDSON, R. G. D.: Trans. Amer. Math. Soc. Vol. 26 (1924) pp. 451–478.

of $|A - \lambda B| = 0$ and are all distinct. Let $A = (a_{rs})$, $B = (b_{rs})$. There exist sets or *poles* $(x_{1k}, \ldots, x_{nk})$, $(y_{l1}, \ldots, y_{ln})$, no one of which consists entirely of zeros, such that

$$\sum_j (a_{ij} - \lambda_k b_{ij})\, x_{jk} = 0\,, \qquad (i, k = 1, 2, \ldots, n)$$

$$\sum_i (a_{ij} - \lambda_l b_{ij})\, y_{li} = 0\,. \qquad (j, l = 1, 2, \ldots, n)$$

These poles may be normalized so that $\Sigma_{i,j}\, y_{ki}\, b_{ij}\, x_{jk} = 1$. Since

$$\sum_{i,j} y_{li}\, a_{ij}\, x_{jk} = \lambda_k \sum_{i,j} y_{li}\, b_{ij}\, x_{jk} = \lambda_l \sum_{i,j} y_{li}\, b_{ij}\, x_{jk}\,,$$

it follows that if $k \neq l$,

$$\sum_{i,j} y_{li}\, a_{ij}\, x_{jk} = 0\,, \qquad \sum_{i,j} y_{li}\, b_{ij}\, x_{jk} = 0\,.$$

The matrices $X = (x_{rs})$ and $Y = (y_{rs})$ are *orthogonal* relative to A and B. Then

$$YBX = I\,, \qquad YAX = [\lambda_1, \lambda_2, \ldots, \lambda_n]\,.$$

In case the roots of $|A - \lambda B| = 0$ are not all distinct, various cases arise. If λ_k is multiple, the number of linearly independent poles $(x_{1k}^{(i)}, \ldots, x_{nk}^{(i)})$ may still be equal to the multiplicity. In this case the argument proceeds as before. The case when there are fewer linearly independent poles than the multiplicity p is called the irregular case. Then the solutions of the equations

$$\sum_j (a_{ij} - \lambda_k b_{ij}) \frac{d^{h-1} x_{jk}}{d\lambda^{h-1}} = \sum_j b_{ij} \frac{d^{h-2} x_{jk}}{d\lambda^{h-2}} \qquad (h = 2, \ldots, p)$$

are used to fill out the rows of X and Y. If λ_k is a root of multiplicity 3, for instance, and if the number of linearly independent poles is 1, the normal form attained for $A - \lambda B$ is

$$\begin{Vmatrix} 0 & 0 & \lambda_k - \lambda \\ 0 & \lambda_k - \lambda & 1 \\ \lambda_k - \lambda & 1 & 0 \end{Vmatrix}.$$

31. Automorphic transformations. If $PAQ = A$, the elements of each matrix being in a commutative field $\mathfrak{F}$, the matrices P and Q determine an *automorphic transformation* of A with respect to the relation of equivalence.

Theorem 31.1. If $d(A) \neq 0$ and M is an arbitrary matrix such that $d(A + M)\, d(A - M) \neq 0$ then

$$P = (A + M)(A - M)^{\mathrm{I}}, \qquad Q = (A + M)^{\mathrm{I}}(A - M)$$

define an automorphic transformation of A. There are no others for which $d(I + P)\, d(I + Q) \neq 0$.[1]

[1] Cayley, A.: Philos. Trans. Roy. Soc. London Vol. 148 (1856) pp. 39–46.

Evidently

$$(I + A^{\mathrm{I}} M)(I - A^{\mathrm{I}} M) = (I - A^{\mathrm{I}} M)(I + A^{\mathrm{I}} M),$$

$$(A + M) A^{\mathrm{I}} (A - M) = (A - M) A^{\mathrm{I}} (A + M).$$

By taking the inverse of each member it follows that

$$(A - M)^{\mathrm{I}} A (A + M)^{\mathrm{I}} = (A + M)^{\mathrm{I}} A (A - M)^{\mathrm{I}},$$

$$(A + M)(A - M)^{\mathrm{I}} A (A + M)^{\mathrm{I}} (A - M) = A.$$

Hence if P and Q both exist, both are non-singular, and they define an automorph of A.

Solving the equations defining P and Q yields, respectively,

$$M = (P + I)^{\mathrm{I}} (P - I) A, \qquad M = A (I - Q)(Q + I)^{\mathrm{I}}.$$

If $PAQ = A$,

$$(P - I) A (Q + I) = (P + I) A (I - Q),$$

and the two values for M are equal. Hence for every P and Q such that $PAQ = A$, and $(P + I)^{\mathrm{I}}$ and $(Q + I)^{\mathrm{I}}$ exist, there is an M in terms of which P and Q may be defined as in the statement of the theorem.

The restriction that $P + I$ and $Q + I$ be non-singular is a serious one, however, and is not easily avoided.

Frobenius[1] noted that if $PAQ = B$, $d(Q) \neq 0$, then $f(P) A [f(Q^{\mathrm{I}})]^{\mathrm{I}} = B$, where f is any rational function such that $f(Q^{\mathrm{I}})$ is non-singular. He also proved that a necessary and sufficient condition in order that P and Q be capable of transforming a non-singular matrix into itself is that it be possible so to order the elementary divisors of $I\lambda - P$ and $I\lambda - Q$ that corresponding elementary divisors are of the same degree and vanish for reciprocal values of λ.

V. Congruence.

32. Matrices with elements in a principal ideal ring. If $A = P^{\mathrm{T}} B P$ where each matrix has elements in a principal ideal ring $\mathfrak{P}$, and if P is unimodular, then A is *congruent* with B, written $A \overset{C}{=} B$. Congruence is an instance of equivalence, and is determinative, reflexive, symmetric, and transitive. (Cf. § 22.)

If a bilinear form $\sum b_{ij} x_i y_j$ of matrix B be transformed by cogredient transformations of matrix P into a form of matrix A, then $A = P^{\mathrm{T}} B P$. It is the purpose of the writer to present the subject as pure matric theory free from the notion of bilinear form, but the reader will have no difficulty in translating the results into the notation of form theory if he so desires. Thus Theorem 34.1 states that

[1] Frobenius: J. reine angew. Math. Vol. 84 (1878) pp. 1–63.

the normal form for a quadratic form is $g_1 x_1^2 + g_2 x_2^2 + \cdots + g_\varrho x_\varrho^2$, and Theorem 32.2 states that the alternating form can be reduced to $h_1(x_1 y_2 - x_2 y_1) + h_2(x_3 y_4 - x_4 y_3) + \cdots$.

Theorem 32.1. If S is symmetric and $S_1 \overset{C}{=} S$, then S_1 is symmetric. If Q is skew and $Q_1 \overset{C}{=} Q$, then Q_1 is skew.

For if $S_1 = P^T S P$, then $S_1^T = P^T S^T P$. If $S^T = S$, then $S_1^T = S_1$. Similarly for Q.

A matrix A with elements in $\mathfrak{P}$ cannot as a rule be written as a sum of a symmetric matrix and a skew matrix (Theorem 5.3) but $2A$ can always be so expressed.

Corollary 32.11. If $S + Q \overset{C}{=} S_1 + Q_1$ where S and S_1 are symmetric and Q and Q_1 are skew, then $S \overset{C}{=} S_1$ and $Q \overset{C}{=} Q_1$.

For by Theorem 5.3 the expression of a matrix as a sum of a symmetric matrix and a skew matrix is unique.

Corollary 32.12. If $A = S + Q$, the invariants of S and Q are invariants of A.

KRONECKER[1] noted that $b - c$ is a cogredience invariant of

$$M = \begin{pmatrix} a & b \\ c & d \end{pmatrix}.$$

Its square is the determinant of the skew component of $2M$.

Theorem 32.2. Every skew matrix Q of rank $\varrho = 2\mu$ is congruent with a direct sum

$$\begin{pmatrix} 0 & h_1 \\ -h_1 & 0 \end{pmatrix} \dot{+} \begin{pmatrix} 0 & h_2 \\ -h_2 & 0 \end{pmatrix} \dot{+} \cdots \dot{+} \begin{pmatrix} 0 & h_\mu \\ -h_\mu & 0 \end{pmatrix} \dot{+} \begin{pmatrix} 0 & 0 \\ 0 & 0 \end{pmatrix} \dot{+} \cdots,$$

where $h_1 | h_2 | \ldots | h_\mu$.[2]

If every element of the first column of

$$Q = \begin{Vmatrix} 0 & q_{12} & q_{13} & \cdots \\ -q_{12} & 0 & q_{23} & \cdots \\ -q_{13} & -q_{23} & 0 & \cdots \\ \cdot & \cdot & \cdot & \cdot \end{Vmatrix}$$

is 0, the columns and corresponding rows may be permuted until this is not so. Let

$$k_{12} = b_2 q_{12} + b_3 q_{13} + \cdots + b_n q_{1n}$$

be a g.c.d. of the elements of the first row. Choose

$$P = \begin{Vmatrix} 1 & 0 & 0 & \cdots \\ 0 & b_2 & * & \cdots \\ 0 & b_3 & * & \cdots \\ \cdot & \cdot & \cdot & \cdot \end{Vmatrix}$$

[1] KRONECKER: Abh. preuß. Akad. Wiss. 1883 II pp. 1–60.

[2] CAHEN, E.: Théorie des nombres Vol. I p. 282. Paris 1914.

unimodular with elements in $\mathfrak{P}$ (Theorem 21.1). Then $P^T QP$ is of the form Q with the further property that $q_{12} \neq 0$ and divides every other element of the first row. By elementary transformations these other elements can be made 0's. The process may be continued until a matrix

$$K = \begin{Vmatrix} 0 & k_{12} & 0 & 0 & \dots \\ -k_{12} & 0 & k_{23} & 0 & \dots \\ 0 & -k_{23} & 0 & k_{34} & \dots \\ 0 & 0 & -k_{34} & 0 & \dots \\ \dots & \dots & \dots & \dots & \dots \end{Vmatrix}$$

is reached.

Either k_{12} divides every other element of K, or another congruent matrix of the same type can be obtained in which the element in the (1, 2)-position has fewer prime factors than k_{12}. By adding row 2, row 3, ..., row n to row 1, and then adding columns similarly, a new congruent matrix is obtained whose first row consists of

$$l_{11} = 0, \quad l_{12} = k_{12} - k_{23}, \quad l_{13} = k_{23} - k_{34}, \quad \dots, \quad l_{1n} = k_{n-1,n}.$$

Every g.c.d. of $(k_{12}, \dots, k_{n-1,n})$ is a g.c.d. of $(l_{12}, \dots, l_{1n})$ and conversely. As in the first part of this proof, a congruent matrix can be obtained similar in form to K but with the element in the (1, 2)-position a g.c.d. of $(k_{12}, \dots, k_{n-1,n})$. Unless $k_{12} | k_{i-1,i}$ for every i, this g.c.d. will have fewer prime factors than k_{12}.

Thus in a finite number of steps a matrix K can be reached in which k_{12} divides every element. By proceeding similarly with the last $n-1$ rows and columns, a matrix K is obtained in which

$$k_{12} | k_{23} | k_{34} | \dots | k_{\varrho-1,\varrho}.$$

Now by adding a proper multiple of row 1 to row 3, k_{23} can be made 0. Every $k_{i,i+1}$ with i even can be made 0 in succession. This proves the theorem.

It is again evident that Q is singular if n is odd (Theorem 8.6). Let $\varrho = 2\mu$.

Theorem 32.3. The numbers $h_1, h_1, h_2, h_2, \dots, h_\mu, h_\mu$ of the canonical form of Q are the invariant factors of Q.

The non-vanishing minor determinants are $\pm h_{k_1} h_{k_2} \dots h_{k_i}$, where each subscript is an integer of the set $1, 2, \dots, \mu$ no integer of which can appear more than twice. It is evident then that the g.c.d. of the i-rowed minor determinants is $d_i = h_1 h_1 h_2 h_2 h_3 \dots$ to i factors. Thus $d_1 = h_1$, $d_2/d_1 = h_1$, $d_3/d_2 = h_2$, $\cdots$, which proves the theorem[1].

Corollary 32.31. The skew canonical form is unique except that the elements may be replaced by associates.

For the invariant factors are invariants (§ 27).

[1] CAHEN: l. c.

Corollary 32.32. Two skew matrices with elements in $\mathfrak{P}$ are congruent if and only if they have the same invariant factors.

Corollary 32.33. If two skew matrices with elements in $\mathfrak{P}$ are equivalent, they are congruent[1].

Corollary 32.34. In a skew matrix the $2i$-th invariant factor is equal to the $(2i-1)$-th[2].

Corollary 32.35. The g.c.d. of the minors of the same even order of a skew matrix is a perfect square[3].

Matrices congruent with each other constitute a *class*.

Corollary 32.36. There is but a finite number of classes of non-singular skew matrices with a given determinant.

Since factorization into primes in a principal ideal ring is unique except for unit factors, there is but a finite number of choices for each invariant factor.

The theory of congruent symmetric matrices is by no means as simple as that of skew matrices. This theory occurs in the literature principally in connection with quadratic forms. The relation $A = P^{\mathrm{T}}BP$ was first given by A. EISENSTEIN[4], who noted that if a quadratic form of matrix B be transformed by a transformation of matrix P, the new quadratic form is of matrix A.

Theorem 32.4. Every symmetric matrix S of rank ϱ with elements in a principal ideal ring $\mathfrak{P}$ is congruent with a matrix of the form

$$\left\|\begin{matrix} s_{11} & s_{12} & 0 & \cdots \\ s_{12} & s_{22} & s_{23} & \cdots \\ 0 & s_{23} & s_{33} & \cdots \\ \cdot & \cdot & \cdot & \cdot \end{matrix}\right\|.$$

where $s_{i-1,i} = s_{ii} = 0$ if $i > \varrho$.

The proof is practically identical with that of the first part of the proof of Theorem 32.2.

This reduced form is not unique, and indeed the problem of finding a unique symmetric canonical form is one of extreme difficulty if it is not actually impossible. It has not been attained even for two-rowed matrices whose elements are rational integers, as will appear in the next section.

33. Matrices with rational integral elements. This topic, which constitutes a large and important chapter in the theory of numbers, can only be touched upon here. Complete references up to the date of their publication are given in articles by K. T. VAHLEN[5], and

[1] FROBENIUS: J. reine angew. Math. Vol. 86 (1879) pp. 146—208.
[2] FROBENIUS: l. c. [3] FROBENIUS: l. c.
[4] EISENSTEIN, A.: J. reine angew. Math. Vol. 35 (1847) pp. 117—136.
[5] VAHLEN, K. T.: Enzykl. math. Wiss. I Vol. 2C2 (1904) pp. 582—638.

L. E. DICKSON[1]. The treatises by P. BACHMANN[2] and DICKSON[3] cover the field quite thoroughly.

The following fundamental theorem was stated by C. HERMITE[4] and proved much later by STOUFF[5].

Theorem 33.1. There is but a finite number of classes of symmetric matrices with rational integral elements of given non-zero determinant. (Cf. Corollary 32.36.)

The proof is too long for inclusion here.

Let

$$A = \begin{Vmatrix} a & b \\ b & c \end{Vmatrix}, \qquad d(A) > 0, \qquad a > 0.$$

Such a matrix is *positive definite.* If

$$-a < 2b \leqq a, \qquad c \geqq a,$$

with $b \geqq 0$ if $c = a$, the matrix A is called *reduced.* Every positive definite symmetric matrix of order 2 is congruent with one and only one reduced matrix[6].

If $d(A) < 0$, A is called *indefinite.* If f is that root of

$$a x^2 + 2bx + c = 0$$

which involves the positive radical, and s is the other root, then A is *reduced* if $|f| < 1$, $|s| > 1$, $fs < 0$. Here again there is at least one reduced form in every class, and usually more than one, but never more than a finite number. By a method of GAUSS these can be arranged into *chains* of reduced forms so that each chain corresponds to one and only one class.

These results are sufficient to indicate the general situation. Canonical forms have been defined in various ways so that every class shall be represented at least once and at most a finite number of times. The goal of defining a unique canonical form has been attained only in special instances.

Theorem 33.2. If $B = (b_{rs})$, where b_{rs} is the positive g.c.d. of r and s, then

$$B \overset{C}{=} [\varphi(1), \varphi(2), \ldots, \varphi(n)],$$

where $\varphi(i)$ is the EULER *φ-function of i.*[7]

Let $\varphi(m)$ be the number of integers in a reduced set of residues modulo m. Then $m = \sum \varphi(d)$ summed over all divisors d of m. Let

1 DICKSON, L. E.: History of the theory of numbers Vol. III pp. 284—288. Washington 1923.

2 BACHMANN, P.: Die Arithmetik der quadratischen Formen. Teubner 1923.

3 DICKSON: Studies in the theory of numbers. Univ. of Chicago Press 1930.

4 HERMITE, C.: J. reine angew. Math. Vol. 47 (1854) p. 336.

5 STOUFF: Ann. École norm. III Vol. 19 (1902) pp. 89—118.

6 KRONECKER: Abh. preuß. Akad. Wiss. 1883 II pp. 1—60.

7 SMITH, H. J. S.: Proc. London Math. Soc. VII 1876 pp. 208—212.

$p_{ij} = 1$ if $i|j$, otherwise $p_{ij} = 0$. Since $p_{ii} = 1$ and $p_{ij} = 0$ for $j < i$, $d(p_{rs}) = 1$. Let $P = (p_{rs})$, $\Phi = [\varphi(1), \varphi(2), \ldots, \varphi(m)]$, $B = (b_{rs}) = P^T \Phi P$. Then

$$b_{rs} = \sum p_{ir} \varphi(i) p_{is} = \sum \varphi(d_{rs})$$

summed over all common divisors d_{rs} of r and s. Hence b_{rs} is the positive g.c.d. of r and s.[1]

34. Matrices with elements in a field. *Theorem 34.1. Every symmetric matrix of rank ϱ with elements in a field $\mathfrak{F}$ not of characteristic 2 is congruent in $\mathfrak{F}$ with a diagonal matrix* $[g_1, g_2, \ldots, g_\varrho, 0, \ldots, 0]$, $g_i \neq 0$.[2]

Consider

$$A = \left\| \begin{matrix} a_{11} & a_{12} & a_{13} & \cdots \\ a_{12} & a_{22} & a_{23} & \cdots \\ a_{13} & a_{23} & a_{33} & \cdots \\ \cdot & \cdot & \cdot & \cdots \end{matrix} \right\| .$$

Assume the minor of order ϱ in principal position (upper left corner) to be non-singular. If $a_{11} = 0$, some $a_{1k} \neq 0$. After adding row k to row 1 and column k to column 1, the new element in the (1, 1)-position is $2a_{1k} \neq 0$, so we assume $a_{11} \neq 0$. Add $-a_{1k}/a_{11}$ times the first row to the k-th row, and similarly for columns, thus reducing all elements of the first row and column to 0 except the first. Now proceed similarly with the lower right minor of order $n-1$, and so on until the diagonal form is reached.

Corollary 34.1. If A_i is the principal minor of order i in the upper left corner of the symmetric matrix A, and if $p_i = d(A_i) \neq 0$, then g_i can be determined as a rational function of the elements of A_i alone.

For in this case none of the first i rows and columns need be interchanged with any of the last $n - i$ rows and columns.

A commutative field $\mathfrak{F}$ is called *ordered* if for every element a of the field one and only one of the relations

$$a = 0, \quad a > 0, \quad -a > 0$$

holds, and if further $a > 0$ and $b > 0$ imply $a + b > 0$ and $ab > 0$.[3]

Theorem 34.2. If $\mathfrak{F}$ is an ordered field, and if $G = [g_1, \ldots, g_\varrho, 0, \ldots, 0]$ is congruent with $H = [h_1, \ldots, h_\varrho, 0, \ldots, 0]$, then the number of g's which are >0 is exactly equal to the number of h's which are >0.

This theorem is SYLVESTERS's[4] "Law of inertia". It was rediscovered by JACOBI[5].

---

[1] Proof by FROBENIUS: J. reine angew. Math. Vol. 86 (1879) pp. 146—208.

[2] For differential forms by LAGRANGE: Misc. Taur. Vol. I (1759) p. 18. — For the general field by L. E. DICKSON: Trans. Amer. Math. Soc. Vol. 7 (1906) pp. 275—292.

[3] VAN DER WAERDEN: Moderne Algebra Vol. I, p. 209.

[4] SYLVESTER: Philos. Mag. IV 1852 pp. 138—142.

[5] JACOBI: J. reine angew. Math. Vol. 53 (1857) pp. 265—270.

Let A and B be any two congruent matrices and suppose that $A = P^{\mathrm{T}}BP$ or

$$a_{rs} = \sum_{i,j} p_{ir} b_{ij} p_{js}.$$

Let $x_1, \ldots, x_n$ be at present undetermined. Then

$$\sum_{r,s} x_r a_{rs} x_s = \sum_{i,j} \Big(\sum_r p_{ir} x_r\Big) b_{ij} \Big(\sum_s p_{js} x_s\Big).$$

Denote $\sum_r p_{ir} x_r$ by y_i. If in particular $A = H$ and $B = G$,

$$\sum_r h_r x_r^2 = \sum_i g_i y_i^2.$$

Now suppose $h_1 > 0, \ldots, h_\varkappa > 0$, $h_{\varkappa+1} < 0, \ldots, h_\varrho < 0$, $g_1 > 0, \ldots, g_\lambda > 0$, $g_{\lambda+1} < 0, \ldots, g_\varrho < 0$, $\lambda < \varkappa$. Then

$$h_1 x_1^2 + \cdots + h_\varkappa x_\varkappa^2 - g_{\lambda+1} y_{\lambda+1}^2 - \cdots - g_\varrho y_\varrho^2$$
$$= g_1 y_1^2 + \cdots + g_\lambda y_\lambda^2 - h_{\varkappa+1} x_{\varkappa+1}^2 - \cdots - h_\varrho x_\varrho^2.$$

Choose $x_{\varkappa+1} = \cdots = x_n = 0$ and $x_1, \ldots, x_\varkappa$ not all zero so that the $\lambda < \varkappa$ linear forms $y_1, \ldots, y_\lambda$ are all zero. Since this implies $h_1 x_1^2 + \cdots + h_\varkappa x_\varkappa^2 = 0$ for the x's not all 0, a contradiction is reached, and it must be true that $\lambda \geqq \varkappa$. Since the relationship between H and G is mutual, $\lambda = \varkappa$.

The number $2\varkappa - \varrho = \sigma$ is called the *signature* of H, and is the number of positive terms diminished by the number of negative terms in the normal form[1]. The two invariants ϱ and σ determine the number of positive and the number of negative terms in the canonical form.

Corollary 34.21. Two symmetric matrices in the real field are congruent if and only if they have the same rank ϱ and the same signature σ.

For every positive h_i can be reduced to 1, and every negative h_i to -1, by an elementary transformation in the real field.

Corollary 34.22. Two symmetric matrices in an algebraically closed field $\mathfrak{C}$ are congruent if and only if they have the same rank ϱ.

For in this case each h_i can be reduced to 1.

A symmetric matrix A in an ordered field is called *positive definite* if $\varrho = \sigma = n$, and *negative definite* if $\varrho = -\sigma = n$. It is *semi-definite* if $\varrho = \sigma$ or $\varrho = -\sigma$.

If A is symmetric and $x_1, \ldots, x_n$ indeterminate, $f = \sum a_{ij} x_i x_j$ is called a *quadratic form*. If $B = P^{\mathrm{T}}AP$, then $f = \sum b_{ij} y_i y_j$ where $x_i = \sum p_{ij} y_j$. If A is positive definite (or negative definite), then $f > 0$ (or $f < 0$) except for $x_1 = x_2 = \cdots = x_n = 0$. If A is positive (or negative) semi-definite, then $f \geqq 0$ ($f \leqq 0$) except for $x_1 = x_2 = \cdots = x_n = 0$. These properties of f characterize A.

A symmetric matrix A of rank ϱ is *regularly arranged* if no two consecutive principal minor determinants $p_1 = a_{11}$, $p_2 = |a_{11}\, a_{22}|$,

[1] FROBENIUS: S.-B. preuß. Akad. Wiss. 1894 I pp. 241–256 and 407–431.

$p_3 = |a_{11}a_{22}a_{33}|, \ldots$ are zero. Such an arrangement is always possible, and if $p_i = 0$, then $p_{i-1}\,p_{i+1} < 0$.[1]

Theorem 34.3. If A is regularly arranged, the signature ϱ of A is equal to the number of permanences minus the number of variations of sign in the sequence

$$1,\ p_1,\ p_2,\ \ldots,\ p_\varrho,$$

where either sign may be attributed to a p which vanishes[2].

It was remarked in Corollary 34.1 that it is possible to reduce A to diagonal form in such a way that $[g_1, \ldots, g_i]$ is obtained from $A_i = \|a_{11} \ldots a_{ii}\|$ by elementary transformations in case the latter is non-singular. That is, $[g_1, \ldots, g_i] = K_i^{\mathrm{T}} A_i K_i$. Hence $g_1 g_2 \ldots g_i$ has the same sign as $p_i = d(A_i)$ in case the latter is not 0, and $g_i < 0$ if and only if p_{i-1} and p_i have opposite signs. In case $p_i = 0$, then p_{i-1} and p_{i+1} have opposite signs, so one permanence and one variation is obtained whether p_i is counted positive or negative. Also $g_1 \ldots g_{i-1}$ and $g_1 \ldots g_{i+1}$ have opposite signs, so one of g_i, g_{i+1} is positive and the other negative. Since σ is invariant, the result is true independently of the method of reduction to normal form.

Corollary 34.3. If A is any matrix, $A^{\mathrm{T}}A$ is positive semi-definite. For

$$p_i = \sum \left(A_{1\,\ldots\,i}^{j_1\,\ldots\,j_i}\right)^2 \geqq 0\,.$$

H. Minkowski[3] proved that two non-singular symmetric matrices with rational elements are congruent in the rational field if and only if three invariants J, A, B coincide. Here J is the number of negative elements in the diagonal form, A is $(-1)^J$ times the product of the primes occuring in the determinant to an odd exponent, and B is a certain product of odd primes. This theory was extended to singular matrices by H. Hasse[4].

L. E. Dickson[5] proved that in the rational field a non-singular symmetric matrix can be reduced to the diagonal form

$$[a, b, c, 1, \ldots, 1, -1, \ldots, -1],$$

where the -1's are absent unless a, b and c are all negative.

Theorem 34.4. Every skew matrix of rank $\varrho = 2\mu$ with elements in a field $\mathfrak{F}$ is congruent with a matrix of the form[6]

$$\begin{pmatrix} 0 & a_{12} \\ -a_{12} & 0 \end{pmatrix} \dot{+} \begin{pmatrix} 0 & a_{34} \\ -a_{34} & 0 \end{pmatrix} \dot{+} \cdots \dot{+} \begin{pmatrix} 0 & a_{\mu-1,\,\mu} \\ -a_{\mu-1,\,\mu} & 0 \end{pmatrix} \dot{+} \begin{pmatrix} 0 & 0 \\ 0 & 0 \end{pmatrix} \dot{+} \cdots.$$

[1] Gundelfinger, S.: J. reine angew. Math. Vol. 91 (1881) pp. 221—237.

[2] Darboux, G.: J. Math. pures appl. II Vol. 19 (1874) pp. 347—396. — Gundelfinger: l. c.

[3] Minkowski, H.: J. reine angew. Math. Vol. 106 (1890) pp. 5—29.

[4] Hasse, H.: J. reine angew. Math. Vol. 152 (1923) pp. 205—224.

[5] Dickson, L. E.: Trans. Amer. Math. Soc. Vol. 7 (1906) pp. 275—292.

[6] Muth, P.: J. reine angew. Math. Vol. 122 (1900) pp. 89—96.

This is a special case of Theorem 32.2.

O. VEBLEN and P. FRANKLIN[1] stated that the analogous form for symmetric matrices does not hold for all fields, but does hold for modular fields.

Corollary 34.4. In the real field every skew matrix of rank $\varrho = 2\mu$ *is congruent with*

$$\begin{pmatrix} 0 & 1 \\ -1 & 0 \end{pmatrix} \dot{+} \cdots \dot{+} \begin{pmatrix} 0 & 1 \\ -1 & 0 \end{pmatrix} \dot{+} \begin{pmatrix} 0 & 0 \\ 0 & 0 \end{pmatrix} \dot{+} \cdots \dot{+} \begin{pmatrix} 0 & 0 \\ 0 & 0 \end{pmatrix},$$

where just μ *blocks are not zero*[2].

Theorem 34.5. If $\mathfrak{F}$ *is a field not of characteristic 2, there exist two n-th order symmetric matrices* A *and* B, $d(B) \neq 0$, *with elements in* $\mathfrak{F}$ *such that* $A + \lambda B$ *has as invariant factors any prescribed polynomials*

$$P_1,\ P_2,\ \ldots,\ P_k$$

with coefficients in $\mathfrak{F}$ *such that* $P_i \mid P_{i+1}$ *and such that the sum of the degrees of the* P*'s is* n.[3]

Let
$$P_i(\lambda) = \lambda^m + b_1 \lambda^{m-1} + b_2 \lambda^{m-2} + \cdots + b_m .$$

Let h be the greatest integer in $\frac{1}{2}(m+1)$, and form the matrix

$$C = \begin{Vmatrix} c_{11} & c_{12} & 0 & 0 & \ldots & 0 & 0 & -\lambda \\ c_{12} & c_{22} & c_{23} & 0 & \ldots & 0 & -\lambda & 1 \\ 0 & c_{23} & c_{33} & c_{34} & \ldots & -\lambda & 1 & 0 \\ \cdot & \cdot & \cdot & \cdot & \cdot & \cdot & \cdot & \cdot \\ 0 & -\lambda & 1 & 0 & \ldots & 0 & 0 & 0 \\ -\lambda & 1 & 0 & 0 & \ldots & 0 & 0 & 0 \end{Vmatrix}$$

of m rows and columns, where the last c is c_{hh}, the c's being for the moment undefined. For m odd, the element in position (h, h) is $c_{hh} - \lambda$, while for m even it is c_{hh}.

For m odd, add λ times column h to column $h-1$, λ times row h to row $h-1$, λ times column $h-1$ to column $h-2$, λ times row $h-1$ to row $h-2$, etc. For m even, start with column $h+1$ and proceed as before. The result in either case is

$$C_1 = \begin{Vmatrix} f(\lambda) & c'_{12} & \ldots & c'_{1h} & 0 & \ldots & 0 \\ c'_{12} & c'_{22} & \ldots & c'_{2h} & 0 & \ldots & 1 \\ \cdot & \cdot & \cdot & \cdot & \cdot & \cdot & \cdot \\ c'_{1h} & c'_{2h} & \ldots & c'_{hh} & \ldots & \ldots & 0 \\ \cdot & \cdot & \cdot & \cdot & \cdot & \cdot & \cdot \\ 0 & 1 & \ldots & 0 & & \ldots & 0 \end{Vmatrix},$$

[1] VEBLEN, O., and P. FRANKLIN: Ann. of Math. II Vol. 23 (1921) pp. 1–15.
[2] BRIOSCHI, F.: J. reine angew. Math. Vol. 52 (1856) pp. 133–141.
[3] DICKSON: Modern algebraic theories, p. 126. Chicago 1926.

where

$$\pm d(C_1) = \pm f(\lambda) = \lambda^m - \sum_{j=1}^{h} c_{jj}\,\lambda^{2j-2} - 2\sum_{j=2}^{h} c_{j-1,j}\,\lambda^{2j-3}.$$

For m odd, $\pm f(\lambda)$ may be identified with $P_i(\lambda)$ by choosing

$$c_{hh} = -b_1, \quad c_{h-1,h} = -\tfrac{1}{2}b_2, \quad \ldots .$$

In case $m = 2h$, the coefficient of λ^{m-1} in $\pm f(\lambda)$ is 0. But if $P_i(\lambda)$ be written

$$P_i(\lambda) = (\lambda - l)^m + b_2'(\lambda - l)^{m-2} + \cdots + b_m',$$

its coefficients may be identified with those of $\pm f(\lambda)$ similarly.

Define A_i to be C with the $-\lambda$'s replaced by 0's for m odd and by l's for m even. Define B_i to have -1's in the secondary diagonal and 0's elsewhere. Then

$$d(A_i + \lambda B_i) = P_i(\lambda), \quad d(B_i) \neq 0.$$

Since the cofactor of a_{11} in $A + \lambda B$ is ± 1, the invariant factors of $A + \lambda B$ are $1, 1, \ldots, P_i(\lambda)$.

Now define

$$A = A_1 \dotplus \cdots \dotplus A_k, \qquad B = B_1 \dotplus \cdots \dotplus B_k.$$

Then $d(B) = \pm 1$, and $A + \lambda B$ has the prescribed invariant factors.

35. Matrices in an algebraically closed field. The results of this section are restricted to matrices with elements in an algebraically closed field $\mathfrak{C}$,[1] due principally to the fact that the following theorem is not valid for a general field, although its analogue for skew matrices (Corollary 32.33) holds for a principal ideal ring.

Theorem 35.1. If A and B are symmetric matrices with elements in $\mathfrak{C}$, and if $A \overset{\mathrm{E}}{=} B$, then $A \overset{\mathrm{C}}{=} B$ in $\mathfrak{C}$.

By Corollary 27.2 $A \overset{\mathrm{E}}{=} B$ if and only if they have the same rank. By Corollary 34.3 $A \overset{\mathrm{C}}{=} B$ if and only if they have the same rank.

Corollary 35.1. Every non-singular symmetric matrix A with elements in $\mathfrak{C}$ can be written $A = R^{\mathrm{T}}R$.

For if A is non-singular, $A \overset{\mathrm{C}}{=} I$ by Corollary 34.3, and $A = R^{\mathrm{T}}IR$.

Theorem 35.1 is not, however, sufficiently explicit to be useful. The following more explicit theorem is due to FROBENIUS[2].

Lemma 35.21.[3] If $f(\lambda)$ is a polynomial of degree $n > 0$, with coefficients in $\mathfrak{C}$, whose constant term is not 0, there exists a polynomial $g(\lambda)$ of degree $< n$ such that

$$[g(\lambda)]^2 \equiv \lambda, \qquad \mathrm{mod}\, f(\lambda).$$

Lemma 35.22. If A is non-singular with elements in $\mathfrak{C}$, there exists a non-singular matrix $X = f(A)$ with elements in $\mathfrak{C}$ such that $X^2 = A$.

---

[1] VAN DER WAERDEN: Moderne Algebra Vol. I p. 198.

[2] FROBENIUS: S.-B. preuß. Akad. Wiss. 1896 pp. 7—16.

[3] Proofs are given in BÔCHER: Introduction to higher algebra, p. 297. — DICKSON: Modern algebraic theories, p. 120.

Let $f(\lambda)$ be the characteristic function of A, and let $g(\lambda)$ be determined as in Lemma 35.21. Let $g(A) = X$.

Theorem 35.2. If $A = PBQ$ where A and B are both symmetric or both skew and P and Q non-singular, then there exists a non-singular matrix R which depends upon P and Q but not upon A and B such that $A = R^T B R$.

If $A = PBQ$ where A and B are both symmetric or both skew, then $A = Q^T B P^T$, and

$$PBQ = Q^T B P^T, \qquad Q^{IT} P B = B P^T Q^I.$$

If U is defined as $Q^{IT} P$, then

$$UB = BU^T, \qquad f(U)B = Bf(U^T)$$

for every polynomial f. Let $f(U) = X$, where $X^2 = U$ (Lemma 35.22). Then

$$R^T B R = Q^T X B X^T Q = Q^T X^2 B Q = Q^T Q^{IT} P B Q = PBQ.$$

Corollary 35.2. If $PAQ = A_1$ and $PBQ = B_1$ where A and A_1, also B and B_1, are both symmetric or both skew, there exists a non-singular matrix R such that

$$R^T A R = A_1, \qquad R^T B R = B_1.$$

Theorem 35.3. If $d(A) \neq 0$, a necessary and sufficient condition that $A \overset{C}{=} B$ in the field $\mathfrak{C}$ is that $\varkappa A + \lambda A^T$ and $\varkappa B + \lambda B^T$ have the same invariant factors[1].

If $A \overset{C}{=} B$, then $A = R^T B R$, and $A^T = R^T B^T R$. Hence

$$\varkappa A + \lambda A^T = R^T(\varkappa B + \lambda B^T) R$$

for $\varkappa$ and λ indeterminate, and $\varkappa A + \lambda A^T$ has the same invariant factors as $\varkappa B + \lambda B^T$ (Theorem 27.2).

If, conversely, $\varkappa A + \lambda A^T$ has the same invariant factors as $\varkappa B + \lambda B^T$, then by Theorem 30.1,

$$A = PBQ, \qquad A^T = PB^T Q.$$

Then

$$A + A^T = P(B + B^T)Q, \qquad A - A^T = P(B - B^T)Q.$$

But $A + A^T$ and $B + B^T$ are symmetric, while $A - A^T$ and $B - B^T$ are skew. Hence there exists a non-singular matrix R such that

$$A + A^T = R^T(B + B^T)R, \qquad A - A^T = R^T(B - B^T)R$$

by Corollary 35.2. Hence, adding[2],

$$A = R^T B R.$$

This theorem is valid also for the case $d(A) = 0$, but the proof is too long for inclusion here[3].

[1] KRONECKER: M.-B. preuß. Akad. Wiss. 1874 pp. 397—447.
[2] Proof by FROBENIUS: S.-B. preuß. Akad. Wiss. 1896 p. 14.
[3] MUTH, P.: Theorie und Anwendung der Elementarteiler, p. 143. Teubner 1899.

Theorem 35.4. Two pairs of matrices (A, B) *and* (A_1, B_1), *where all are symmetric, or all are skew, or* A, A_1 *are symmetric and* B, B_1 *are skew, are congruent if and only if they are equivalent.*

This follows directly from Corollary 35.2.

P. MUTH[1] proved that two congruent pairs of real symmetric matrices A, B and A_1, B_1 such that $\varkappa A + \lambda B$ has only imaginary elementary divisors, are congruent with respect to the real field. If $\varkappa A + \lambda B \overset{C}{=} \varkappa A_1 + \lambda B_1$, and if B and B_1 are semi-definite of the same sign, the pencils are congruent with respect to the real field if they have no linear elementary divisors with basis $\varkappa$; otherwise if and only if A and A_1 have the same signature.

L. E. DICKSON[2] gave necessary and sufficient conditions for the congruence of pairs of symmetric matrices, both singular and non-singular, with respect to the real field. Also[3] he gave conditions for the congruence of pairs of two-rowed symmetric matrices in any field.

36. Hermitian matrices. This theory is an abstraction of the theory of hermitian forms $\sum a_{ij} x_i \bar{x}_j$ $(a_{ij} = \bar{a}_{ji})$ under conjunctive transformations, and the theorems may be so interpreted. Let $\mathfrak{F}$ be any field, and k any number of $\mathfrak{F}$ which is not a square in $\mathfrak{F}$. The field $\mathfrak{F}(\xi_1) = \mathfrak{H}$ obtained by adjoining to $\mathfrak{F}$ a root ξ_1 of the equation $\xi^2 = k$ is identical with its conjugate field $\mathfrak{F}(\xi_2)$ obtained by adjoining to $\mathfrak{F}$ the other root ξ_2. Thus the substitution of ξ_2 for ξ_1 defines an automorphism of the field $\mathfrak{H}$. If h is a number of $\mathfrak{H}$, we shall denote by $\bar{h}$ the number of $\mathfrak{H}$ to which h corresponds under this automorphism. Evidently $\bar{h} = h$ if and only if h is in $\mathfrak{F}$.

Let $A = (a_{rs})$ be any matrix with elements in H. The conjugate of A is by definition

$$A^{\mathrm{C}} = (\bar{a}_{rs}).$$

If $A^{\mathrm{T}} = A^{\mathrm{C}}$, A is called *hermitian*[4]. The diagonal elements of an hermitian matrix are in $\mathfrak{F}$.

If $A^{\mathrm{T}} = -A^{\mathrm{C}}$, A is called *skew hermitian.* (Cf. § 18.) The diagonal elements of a skew hermitian matrix are multiples of ξ_1 by a number of $\mathfrak{F}$.

Some properties of the characteristic roots of an hermitian matrix when $\mathfrak{H}$ is the complex field were developed in § 18.

Two matrices A and B are *conjunctive* or *congruent in the hermitian sense* (written $A \overset{H}{=} B$) if and only if there exists a non-singular matrix P with elements in $\mathfrak{H}$ such that

$$A = P^{\mathrm{T}} B P^{\mathrm{C}}.$$

[1] MUTH, P.: J. reine angew. Math. Vol. 128 (1905) pp. 302—321.
[2] DICKSON, L. E.: Trans. Amer. Math. Soc. Vol. 10 (1909) pp. 347—360.
[3] DICKSON, L. E.: Amer. J. Math. Vol. 31 (1909) pp. 103—108.
[4] See C. HERMITE: C. R. Acad. Sci., Paris Vol. 41 (1855) p. 181.

This relationship is determinative, reflexive, symmetric and transitive. If A, B and P have elements in $\mathfrak{F}$, $A \overset{H}{=} B$ becomes $A \overset{C}{=} B$.

Theorem 36.1. If A is hermitian (skew hermitian) and $B \overset{H}{=} A$, then B is hermitian (skew hermitian).

For $B = P^{T} A P^{C}$ implies

$$B^{C} = P^{CT} A^{C} P, \qquad B^{CT} = P^{T} A^{CT} P^{C}, \qquad P^{T} A P^{C} = B,$$

and similarly for the skew hermitian case.

There is a marked parallelism between the properties of hermitian matrices under conjunctive transformations and symmetric matrices under cogredient transformations. The proofs also are parallel and will be omitted.

Theorem 36.2. Every hermitian matrix of rank ϱ with elements in a field $\mathfrak{H}$ without characteristic 2 is conjunctive in $\mathfrak{H}$ with a diagonal matrix $[g_1, g_2, \ldots, g_\varrho, 0, \ldots, 0]$, g_i in $\mathfrak{F}$ and $\neq 0$.

The proof is similar to that of Theorem 34.1.

Theorem 36.3. If $\mathfrak{F}$ is an ordered field and $\xi^2 < 0$, and if $G = [g_1, \ldots, g_\varrho, 0, \ldots, 0]$ is conjunctive with $H = [h_1, \ldots, h_\varrho, 0, \ldots, 0]$ in the field $\mathfrak{H} = \mathfrak{F}(\xi)$, then the number of g's which are > 0 is exactly equal to the number of h's which are > 0.

For if $\xi^2 < 0$, $x\bar{x} \geqq 0$, and $x\bar{x} = 0$ if and only if $x = 0$. The proof proceeds as in the proof of Theorem 34.2 with x_r^2 replaced by $x_r \bar{x}_r$, etc.

If $\mathfrak{F}$ is the real field so that $\mathfrak{H}$ is the complex field, then each element of the diagonal form can be reduced to $+1$ or -1. Hence

Theorem 36.4. Two hermitian matrices are conjunctive in the complex field $\mathfrak{C}$ if and only if they have the same rank and the same signature.

Theorem 36.5. A necessary and sufficient condition in order that two pairs of hermitian matrices A, B and A_1, B_1, A and A_1 non-singular, be conjunctive in $\mathfrak{C}$ is that $\lambda A + B$ and $\lambda A_1 + B_1$ have the same invariant factors.

Theorem 36.6. There exist pairs of hermitian matrices of order n, one of which is non-singular, having any given admissible elementary divisors.

Only minor changes in the proofs of the corresponding theorems for symmetric matrices is necessary to obtain proofs of the above theorems[1]. In fact symmetric and hermitian matrices are considered simultaneously by L. E. DICKSON[2].

If λ is a characteristic root of $A = (a_{rs})$, there exist n numbers $(x_1, \ldots, x_n)$ in the field of the elements of A such that

$$\sum_j (a_{ij} - \lambda \delta_{ij}) x_j = 0. \qquad (i = 1, 2, \ldots, n)$$

---

[1] LOGSDON, M. I.: Amer. J. Math. Vol. 44 (1922) pp. 254–260.

[2] DICKSON, L. E.: Modern algebraic theories, Chap. IV. Chicago 1926.

These numbers $x_1, \ldots, x_n$ constitute a *pole* of A corresponding to the characteristic root λ. They are defined only up to a non-zero factor. If $\Sigma_i x_i \bar{x}_i = d^2$, then $(x_1/d, x_2/d, \ldots, x_n/d)$ is a *normalized pole.* If X is a matrix whose i-th row is a normalized pole of A corresponding to the characteristic root λ_i, then X is a *polar matrix* of A.

Theorem 36.7. If A is hermitian (symmetric) with distinct characteristic roots $\lambda_1, \lambda_2, \ldots, \lambda_n$, its polar matrix is unitary (orthogonal)[1].

Let $x_1, \ldots, x_n, y_1, \ldots, y_n$ be any numbers of $\mathfrak{H}$ and define

$$\xi_i = \sum a_{ij} x_j, \qquad \eta_i = \sum a_{ij} y_j.$$

If A is hermitian,

$$\sum \xi_i \bar{y}_i = \sum a_{ij} x_j \bar{y}_i = \sum \bar{a}_{ji} \bar{y}_i x_j = \sum \bar{\eta}_j x_j.$$

If in particular $(x_1, \ldots, x_n) = (x_{1p}, \ldots, x_{np})$ is a normalized pole corresponding to λ_p and $(y_1, \ldots, y_n) = (x_{1q}, \ldots, x_{nq})$ is a normalized pole corresponding to λ_q,

$$\xi_i = \lambda_p x_{ip}, \quad \eta_i = \lambda_q x_{iq},$$

$$\sum_i \lambda_p x_{ip} \bar{x}_{iq} = \sum_j \lambda_q x_{jq} \bar{x}_{jq}$$

since the λ's are real. Then

$$(\lambda_p - \lambda_q) \sum_i x_{ip} \bar{x}_{iq} = 0.$$

If $\lambda_p \neq \lambda_q$, $\Sigma_i x_{ip} \bar{x}_{iq} = 0$, and if $p = q$, $\Sigma_i x_{ip} \bar{x}_{iq} = 1$. Hence X is unitary, and if it is real, it is orthogonal[2].

Corollary 36.7. If A is hermitian with distinct characteristic roots λ_i, and if X is its polar matrix, then

$$X^{\mathrm{T}} A X = D = [\lambda_1, \lambda_2, \ldots, \lambda_n].$$

For

$$\sum_j (a_{ij} - \lambda_k \delta_{ij}) x_{jk} = 0$$

may be written $AX = XD$. Then

$$X^{\mathrm{T}} A X = X^{\mathrm{T}} X D = D$$

since X is unitary.

Theorem 36.8. The elementary divisors of the characteristic matrix of every hermitian matrix are simple[3].

This follows immediately from Theorem 18.6.

If a matrix is of order n and signature σ, its *characteristic* is $q = \frac{1}{2}(n - |\sigma|)$. [Loewy.]

[1] Laurent, H.: Nouv. Ann. III Vol. 16 (1897) pp. 149—168.

[2] Buchheim, A.: Mess. Math. Vol. 14 (1885) pp. 143—144.

[3] Christoffel, E. B.: J. reine angew. Math. Vol. 63 (1864) pp. 255—272. — Autonne, L.: Bull. Soc. Math. France Vol. 31 (1903) pp. 268—271. — Baker, H. F.: Proc. London Math. Soc. Vol. 35 (1903) pp. 379—384.

Theorem 36.9. If H and K are hermitian with $d(H) \neq 0$, *and if* q *is the characteristic of* K, *then*

$$q \geqq s + \sum \left[\frac{h}{2}\right] + \sum \left[\frac{h'-1}{2}\right],$$

where $2s$ *is the sum of the exponents of the elementary divisors of* $\varrho H - K$ *which vanish for imaginary values of* ϱ, h *varies over the exponents of those elementary divisors which vanish for real non-zero values of* ϱ, *and* h' *varies over those which vanish for* $\varrho = 0$. $[k]$ *signifies the greatest integer in* k.[1]

The proof is too long for inclusion here. T. J. I'A. BROMWICH[2] gave a proof for the real case, and extended the theorem to include the cases $d(H) = 0$ and $d(\varrho H - K) \equiv 0$.

Corollary 36.91. If H and K are hermitian, H non-singular of signature σ, *then* $d(\varrho H - K) = 0$ *has at least* $|\sigma|$ *real roots*[3].

For LOEWY's inequality gives $n - |\sigma| \geqq 2s$ or $n - 2s \geqq |\sigma|$.

Corollary 36.92. If H and K are hermitian, H positive definite, then $d(\varrho H - K) = 0$ *has only real roots*[4].

This is a special case of KLEIN's theorem. Corollary 18.31 is in turn a special case of the last corollary.

It had been shown by SYLVESTER[5] that the number of real roots of $d(\varrho H - K) = 0$ was $\geqq$ the signature of every matrix of the family, H and K symmetric.

37. Automorphs. If $P^{CT}AP = A$, then P is a *conjunctive automorph* of A. A unitary matrix may be considered as an automorph of I. If $P^{T}AP = A$, P is a *cogredient automorph* of A. If $A = I$, P is orthogonal.

L. AUTONNE[6] called *lorenzian* any real automorph of a non-singular real symmetric matrix A. Previously LAUE[7] and A. BRILL[8] had called a matrix lorenzian if it was a cogredient automorph of the diagonal matrix $[1, 1, 1, -1]$.

Theorem 37.1. If A is symmetric and non-singular with elements in a field $\mathfrak{F}$, *and if Q is an arbitrary skew matrix such that* $d(A+Q)(A-Q) \neq 0$, *then*

$$P = (A+Q)^{I}(A-Q)$$

---

1 LOEWY, A.: J. reine angew. Math. Vol. 122 (1900) pp. 53–72 — Nachr. Ges. Wiss. Göttingen 1900 pp. 298–302.

2 BROMWICH, T. J. I'A.: Proc. London Math. Soc. I Vol. 32 (1900) pp. 321 to 352.

3 KLEIN, F.: Dissertation. Bonn 1868 — Math. Ann Vol. 23 (1884) pp. 539 to 578.

4 CHRISTOFFEL, E. B.: J. reine angew. Math. Vol. 63 (1864) pp. 255–272.

5 SYLVESTER: Philos. Mag. Vol. 6 (1853) pp. 214–216.

6 AUTONNE, L.: C. R. Acad. Sci., Paris Vol. 156 (1913) pp. 858–860 — Ann. Univ. Lyon II Vol. 38 (1915) pp. 1–77.

7 LAUE: Das Relativitätsprinzip. Vieweg 1911.

8 BRILL, A.: Das Relativitätsprinzip. Teubner 1912.

is a cogredient automorph of A.[1] *If* $d(I+P) \neq 0$, *there are no others*[2].

For
$$\begin{aligned} P^{\mathrm{T}} &= (A-Q)^{\mathrm{T}}(A+Q)^{\mathrm{I}\,\mathrm{T}} \\ &= (A^{\mathrm{T}}-Q^{\mathrm{T}})(A^{\mathrm{T}}+Q^{\mathrm{T}})^{\mathrm{I}} \\ &= (A+Q)(A-Q)^{\mathrm{I}}. \end{aligned}$$

That $P^{\mathrm{T}}AP = A$ now follows from Theorem 31.1.

If the equation defining P be solved for Q, there results

$$Q = A\frac{I-P}{I+P}.$$

This expression exists if $d(I+P) \neq 0$, and is well-defined by Theorem 15.6. It remains to be shown that Q is skew if $P^{\mathrm{T}}AP = A$.

$$\begin{aligned} Q^{\mathrm{T}} &= \left(\frac{I-P}{I+P}\right)^{\mathrm{T}} A^{\mathrm{T}} = \frac{I-P^{\mathrm{T}}}{I+P^{\mathrm{T}}}A \\ &= \frac{I-A\,P^{\mathrm{I}}A^{\mathrm{I}}}{I+A\,P^{\mathrm{I}}A^{\mathrm{I}}}A = A\frac{I-P^{\mathrm{I}}}{I+P^{\mathrm{I}}} \\ &= A\frac{P-I}{P+I} = -Q. \end{aligned}$$

Many attempts have been made to remove the restriction $d(I+P) \neq 0$, which is not trivial. H. Taber[3] showed that such an automorph can be represented as the product of two automorphs of the Cayley type.

A. Voss[4] extended the above theorem to a non-singular A not necessarily symmetric as follows: Every P such that $P^{\mathrm{T}}AP = A$ for which $d(P+\eta I) \neq 0$, $\eta = \pm 1$, can be uniquely expressed in the form

$$P = \eta(I-BA)(I+BA)^{\mathrm{I}},$$

where $BA + B^{\mathrm{T}}A^{\mathrm{T}} = 0$. He also found the number of linearly independent solutions B of this latter equation, and proved that the number of parameters in P for A non-singular is $m-\mu$ where m is the number of linearly independent matrices C such that $AC^{\mathrm{T}} = CA$. He showed later[5] that $m-\mu = \frac{1}{2}n(n-1) - \frac{1}{2}(n-\varrho)(n-\varrho-1)$ where A is of order n and rank ϱ.

A. Loewy[6] showed that if $d(P+\eta I) = 0$, two transformations of the type given by Voss will generate P.

J. H. M. Wedderburn[7] used the exponential function of a matrix to obtain the general solution X of $X^{\mathrm{T}}PX = P$ for P non-singular, the parameters entering transcendentally.

---

1 Cayley: Philos. Trans. Roy. Soc. London Vol. 148 (1856) pp. 39–46.
2 Frobenius: J. reine angew. Math. Vol. 84 (1878) pp. 1–63.
3 Taber, H.: Math. Ann. Vol. 46 (1895) pp. 561–583.
4 Voss, A : Abh. bayer. Akad. Wiss. II Vol. 17 (1892) pp. 235–356.
5 Voss, A.: Abh. bayer. Akad. Wiss. Vol. 26 (1896) pp. 1–23.
6 Loewy, A.: Math. Ann. Vol. 48 (1897) pp. 97–110.
7 Wedderburn, J. H. M.: Ann. of Math. II Vol. 23 (1921) pp. 122–134.

HERMITE[1] proved that if A is symmetric and non-singular with rational integral elements, every automorph P with rational integral elements is of the form

$$P_1^{e_1} P_2^{e_2} \dots P_k^{e_k},$$

where k is finite and the P's commutative, and the exponents are positive or negative integers or zero.

H. POINCARÉ[2] stated without proof that if A has rational integral elements and $ASA^{\mathrm{T}} = S$ where S is the skew normal form of BRIOSCHI (Corollary 34.4), then A is a product of elementary matrices of two simple types. This was proved by H. R. BRAHANA[3] and extended to the case where S is not in normal form.

Theorem 37.2. In the complex field a necessary and sufficient condition that a matrix P be a congruent automorph of some non-singular matrix is that $P = AB$ where A and B are involutory[4].

It will first be shown that there exists a non-singular matrix H such that $PHP = H$ if and only if $P = AB$ where A and B are involutory. Evidently

$$PAP = ABA^2B = AB^2 = A$$

so the condition is sufficient. If $PHP = H$, then $PHPHP = H^2P$ and $PH^2 = H^2P$, so H^2 is commutative with P. Then $Pf(H^2) = f(H^2)P$ for every rational function f. Choose f so that $f(H^2) = K$ (Lemma 35.22) where $K^2 = H^2$. Define $A = HK^{\mathrm{I}}$. Since K^{I} is a polynomial in H, $A = K^{\mathrm{I}}H$. Then $A^2 = H^2H^{-2} = I$ so that A is involutory. Define $B = AP = A^{\mathrm{I}}P$ so that $P = AB$. Then $B^2 = APA^{\mathrm{I}}P = K^{\mathrm{I}}HPH^{\mathrm{I}}KP = K^{\mathrm{I}}P^{\mathrm{I}}KP = K^{\mathrm{I}}P^{\mathrm{I}}PK = I$, so B also is involutory.

It will be shown, secondly, that there exists a non-singular matrix L such that $PLP^{\mathrm{T}} = L$ if and only if there exists a non-singular H such that $PHP = H$. Since for any M, $M - \lambda I$ and $M^{\mathrm{T}} \quad \lambda I$ have the same invariant factors, the pairs (M, M^{I}) and (I, I) are equivalent (Theorem 30.2). Thus there exist matrices R and S such that

$$RMS = M^{\mathrm{T}}, \qquad RIS = I.$$

Hence $RMR^{\mathrm{I}} = M^{\mathrm{T}}$. Take $M = P^{\mathrm{I}\,\mathrm{T}}$ so that

$$H^{\mathrm{I}}PH = P^{\mathrm{I}} = RP^{\mathrm{I}\,\mathrm{T}}R^{\mathrm{I}},$$

$$R^{\mathrm{I}}H^{\mathrm{I}}PHR = P^{\mathrm{I}\,\mathrm{T}},$$

$$P(HR)P^{\mathrm{T}} = HR.$$

The converse of this step follows similarly[5].

---

[1] HERMITE: J. reine angew. Math. Vol. 47 (1854) pp. 307—368.

[2] POINCARÉ, H.: Rend. Circ. mat. Palermo Vol. 18 (1904) pp. 45—110.

[3] BRAHANA, H. R.: Ann. of Math. II Vol. 24 (1923) pp. 265—270.

[4] JACKSON, D.: Trans. Amer. Math. Soc. Vol. 10 (1909) pp. 479—484.

[5] Proof by FROBENIUS: S.-B. preuß. Akad. Wiss. 1910 I pp. 3—15.

P. F. SMITH[1] had previously shown that a cogredient automorph is a product of not more than n involutory matrices.

H. HILTON[2] called *quasi-unitary* a conjunctive automorph of a non-definite canonical hermitian matrix.

L. AUTONNE gave a systematic treatment of lorenzian matrices[3]. A necessary and sufficient condition that A be lorenzian is that it be an automorph of a diagonal matrix whose diagonal elements are $+1$ or -1. The most general lorenzian is of the form

$$K = LFM, \qquad F = I_{u-v} \dot{+} K \dot{+} I_{n-u-v},$$

where L and M are direct sums of orthogonal matrices, and

$$K = \begin{Vmatrix} T & H \\ H & T \end{Vmatrix}, \qquad T^2 - H^2 = I_v,$$

T and H being canonical hermitians. For $n = 4$ he obtained BRILL's canonical form[4]:

$$F = \begin{Vmatrix} 1 & 0 & 0 & 0 \\ 0 & 1 & 0 & 0 \\ 0 & 0 & \theta & \eta \\ 0 & 0 & \eta & \theta \end{Vmatrix}, \qquad \theta^2 - \eta^2 = 1,$$

where $\theta = k$, $\eta = kq$, $k = 1/\sqrt{1 - q^2}$.

VI. Similarity.

38. Similar matrices. Two matrices A and B with elements in a principal ideal ring $\mathfrak{P}$ are called *similar* (written $A \equiv B$) if there exists a unimodular matrix P such that $A = P^{\mathrm{I}} B P$.[5] Similarity is an instance of equivalence, and is determinative, reflexive, symmetric and transitive (§ 22). More than this, every unimodular matrix P determines an automorphism of the ring of matrices with elements in $\mathfrak{P}$, for if

$$A_1 = P^{\mathrm{I}} B_1 P, \qquad A_2 = P^{\mathrm{I}} B_2 P,$$

then

$$A_1 + A_2 = P^{\mathrm{I}} (B_1 + B_2) P, \qquad A_1 A_2 = P^{\mathrm{I}} (B_1 B_2) P.$$

A matrix may be interpreted as a linear homogeneous transformation in vector space. From this point of view similar matrices represent the same transformation referred to different bases. All the theorems of this chapter may be interpreted from this standpoint.

---

[1] SMITH, P. F.: Trans. Amer. Math. Soc. Vol. 6 (1905) pp. 1–16.

[2] HILTON, H.: Ann. of Math. II Vol. 15 (1914) pp. 195–201.

[3] AUTONNE, L.: Ann. Univ. Lyon II Vol. 38 (1915) pp. 1–77.

[4] BRILL: l. c. p. 31.

[5] FROBENIUS: J. reine angew. Math. Vol. 84 (1878) p. 21.

Theorem 38.1. The coefficients of the characteristic equation of a matrix A are similarity invariants[1].

For if $A = P^{\mathrm{I}}BP$, then $A - \lambda I = P^{\mathrm{I}}(B - \lambda I)P$, and $d(A - \lambda I) = d(B - \lambda I)$.

Corollary 38.1. If $A \overset{S}{=} B$ in an algebraically closed field, the characteristic roots of A coincide with those of B, and each has the same multiplicity for A as for B.

The number theory of similar matrices has received relatively little attention as compared with the number theory of congruent matrices. What has been done has been mainly concerned with linear transformations and groups.

C. JORDAN[2] defined a canonical form using integral algebraic numbers, and gave a necessary and sufficient condition that two such matrices be commutative.

L. E. DICKSON[3] generalized to GALOIS fields the canonical form of JORDAN. He also gave[4] an explicit form of all m-ary linear homogeneous substitutions in $GF(p^n)$ commutative with a particular one.

39. Matrices with elements in a field. *Theorem 39.1. A necessary and sufficient condition that two matrices A and B with elements in a field $\mathfrak{F}$ be similar is that, in the polynomial domain $\mathfrak{F}(\lambda)$, $I\lambda - A$ and $I\lambda - B$ have the same invariant factors.*

If $A = P^{\mathrm{I}}BP$, then evidently $I\lambda - A = P^{\mathrm{I}}(I\lambda - B)P$, so that $I\lambda - A$ and $I\lambda - B$ have the same invariant factors (Theorem 27.2).

If, conversely, $I\lambda - A$ and $I\lambda - B$ have the same invariant factors, there exist two non-singular matrices Q and P whose elements are independent of λ such that

$$I\lambda - A = Q(I\lambda - B)P$$

by Theorem 30.1. Hence

$$I = QP, \quad Q = P^{\mathrm{I}}, \quad A = P^{\mathrm{I}}BP.$$

Corollary 39.11. In an algebraically closed field,

$$A \overset{S}{=} J = J_n \dot{+} J_{n-1} \dot{+} \cdots \dot{+} J_{n-k},$$

where J_i is the JORDAN *matrix of the i-th invariant factor of A.*[5]

Cf. Corollary 30.22.

---

[1] FUCHS, L.: J. reine angew. Math. Vol. 66 (1866) pp. 121–160.

[2] JORDAN, C.: Traité des substitutions et des équations algébriques, p. 125. Paris 1870.

[3] DICKSON, L. E.: Amer. J. Math. Vol. 22 (1900) pp. 121–137.

[4] DICKSON, L. E.: Proc. London Math. Soc. Vol. 32 (1900) pp. 165–170.

[5] JORDAN, C.: Traité des substitutions et des équations algébriques, p. 114. Paris 1870.

Corollary 39.12. In any field,

$$A \overset{S}{=} B = B_n \dotplus B_{n-1} \dotplus \cdots \dotplus B_{n-k},$$

where B_i is the companion matrix of the i-th invariant factor of A.

Cf. Corollary 30.21.

WEIERSTRASS[1] noted that A can be reduced to diagonal form if and only if its elementary divisors are simple.

Various methods of reduction to the JORDAN form have been given by E. NETTO[2], H. HILTON[3], and G. VOGHERA[4].

The second form B was introduced by FROBENIUS[5] for the complex field. Derivations of the normal form (which, however, introduce irrationalities in the derivation) were given by G. LANDSBERG[6], W. BURNSIDE[7], and H. HILTON[8].

A number of writers have given *a priori* proofs of Corollary 39.12 and have used this as a starting point in the development of the entire theory of matrices. The first complete proof valid for a general field, but restricted to non-singular matrices, was given by S. LATTÈS[9]. The method is partly geometric, having been suggested, the author states, by a paper of C. SEGRE[10] and a book by E. BERTINI[11]. If $A = (a_{rs})$ is non-derogatory, it has a finite number of poles, and $\alpha_1, \alpha_2, \ldots, \alpha_n$ are so chosen that the hyperplane

$$y_1 = \alpha_1 x_1 + \alpha_2 x_2 + \cdots + \alpha_n x_n = 0$$

does not contain a pole of A. Let $(\beta_{rs}^{(i)}) = A^i$ and set $P = \Sigma_i \alpha_i \beta_{is}^{(r-1)}$. Then

$$P^{\mathrm{I}} A P = \left\| \begin{matrix} 0 & 1 & 0 & \ldots & 0 \\ 0 & 0 & 1 & \ldots & 0 \\ \cdot & \cdot & \cdot & \cdot & \cdot \\ -a_n & -a_{n-1} & -a_{n-2} & & -a_1 \end{matrix} \right\|.$$

If A has an infinite number of poles—i.e., is derogatory—the rows of P are not independent. If the first m rows are independent, A can be reduced to the form

$$\left\| \begin{matrix} B & O \\ C & D \end{matrix} \right\|, \quad B = \left\| \begin{matrix} 0 & 1 & \ldots & 0 \\ \cdot & \cdot & \cdot & \cdot \\ -a_m & -a_{m-1} & \ldots & -a_1 \end{matrix} \right\|.$$

1 WEIERSTRASS: M.-B. preuß. Akad. Wiss. 1868 pp. 310–338.
2 NETTO, E.: Acta math. Vol. 17 (1893) pp. 265–280.
3 HILTON, H.: Mess. of Math. Vol. 39 (1909) pp. 24–26.
4 VOGHERA, G.: Boll. Un. Mat. Ital. Vol. 7 (1928) pp. 32–34.
5 FROBENIUS: J. reine angew. Math. Vol. 86 (1879) pp. 146–208.
6 LANDSBERG, G.: J. reine angew. Math. Vol. 116 (1896) pp. 331–349.
7 BURNSIDE, W.: Proc. London Math. Soc. Vol. 30 (1898) pp. 180–194.
8 HILTON, H.: Homogeneous linear substitutions. Oxford 1914.
9 LATTÈS, S.: Ann. Fac. Sci. Univ. Toulouse Vol. 28 (1914) pp. 1–84.
10 SEGRE, C.: Atti Accad. naz. Lincei, Mem., III Vol. 19 (1884) pp. 127–148.
11 BERTINI, E.: Introduzione alla geometra proiettiva degli iparspazi. Pisa 1907.

Then by a proper choice of the α's C can be made O, and the reduction continued until

$$A \overset{S}{=} B_1 \dotplus B_2 \dotplus \cdots \dotplus B_p,$$

each B being of the form of B above, and such that $|B_i - \lambda I|$ divides $|B_{i+1} - \lambda I|$. The form is shown to be unique.

G. Kowalewski[1] sketched an alternative treatment of the same problem, stating that the investigation had been prompted by a remark of Sophus Lie that such a reduction would be desirable. Let the point $(x_1, x_2, \ldots, x_n)$ be denoted by (x), and write

$$x_i' = \sum l_{ij} x_j, \qquad L = (l_{rs})$$

in the notation $(x') = (x) L$. If $f(\omega)$ is a polynomial of degree m, denote

$$(x) f(L) = (y) = k_0(x) + k_1(x) L + \cdots + k_m(x) L^m.$$

Let (ξ) be a definite point. Let α be such that (ξ), $(\xi) L$, $(\xi) L^2, \ldots$, $(\xi) L^{\alpha-1}$ are linearly independent, while

$$(\xi) L^\alpha = a_0(\xi) + a_1(\xi) L + \cdots + a_{\alpha-1}(\xi) L^{\alpha-1}.$$

Choose (ξ) so that α is as large as possible. Then

$$A(\omega) = a_0 + a_1 \omega + \cdots + a_{\alpha-1} \omega^{\alpha-1} - \omega^\alpha$$

is called the *first characteristic polynomial* of L. In case $\alpha < n$, take (η) a point not dependent upon (ξ), $(\xi) L, \ldots, (\xi) L^{\alpha-1}$ and choose β so that

$$(\eta) L^\beta = b_0(\eta) + b_1(\eta) L + \cdots + b_{\beta-1}(\eta) L^{\beta-1} + (\xi) P(L)$$

for some polynomial P, while no such relation holds for a smaller β. Take (η) so that β is maximal. If

$$B(\omega) \equiv b_0 + b_1 \omega + \cdots + b_{\beta-1} \omega^{\beta-1} - \omega^\beta,$$

then

$$(\eta) B(L) + (\xi) P(L) = 0.$$

Call B a *second characteristic polynomial.* If $\alpha + \beta < n$, continue and get

$$(\zeta) C(L) + (\eta) R(L) + (\xi) Q(L) = 0.$$

Continue until $\alpha + \beta + \gamma + \cdots = n$. It is now possible to take new points

$$(\eta') = (\eta) + (\xi) \varphi(L),$$

$$(\zeta') = (\zeta) + (\eta) \psi(L) + (\xi) \chi(L), \qquad \text{etc.}$$

so that

$$(\xi) A(L) = (\eta') B(L) = (\zeta') C(L) = \cdots = 0.$$

Then $B|A, C|B, \ldots$ Now transform

$$(\xi),\quad (\xi) L,\quad \ldots,\quad (\xi) L^{\alpha-1},\quad (\eta'),\quad (\eta') L,\quad \ldots,\quad (\eta') L^{\beta-1},\quad \ldots$$

---

[1] Kowalewski, G.: Ber. Verh. sächs. Akad. Leipzig Vol. 68 (1916) pp. 325 to 335.

into $(1, 0, \ldots, 0), (0, 1, \ldots, 0), \ldots$ respectively. Then the transform of L is a direct sum of blocks of the type

$$\begin{Vmatrix} 0 & 0 & \ldots & 0 & a_0 \\ 1 & 0 & \ldots & 0 & a_1 \\ \cdot & \cdot & \cdot & \cdot & \cdot \\ 0 & 0 & \ldots & 1 & a_k \end{Vmatrix}.$$

Another treatment of the same problem was given by W. KRULL[1] at the suggestion of A. LOEWY. An arbitrary matrix A is similar to one of the form

$$\begin{Vmatrix} A_1^* & 0 \\ B & A_2 \end{Vmatrix},$$

where A_1^* is the companion matrix of the minimum equation of A. The above matrix is shown to be similar to

$$\begin{Vmatrix} A_1^* & 0 \\ 0 & A_2 \end{Vmatrix}.$$

The process is continued with A_2 until it is shown that

$$A \overset{S}{=} A_1^* \dotplus A_2^* \dotplus \cdots \dotplus A_m^*,$$

where $|I\lambda - A_i^*| = 0$ is the minimum equation of $A_i^* \dotplus \cdots \dotplus A_m^*$. The form is shown to be unique.

L. E. DICKSON[2] gave an independent development which is comparatively brief. Instead of the planes used by LATTÈS, he used polynomial chains, and the restriction $d(A) \neq 0$ is removed.

A. A. BENNETT[3] discussed the computational aspects of the methods of LATTÈS, KOWALEWSKI and DICKSON.

A clear presentation and refinement of the method of KOWALEWSKI was given by H. W. TURNBULL and A. C. AITKEN[4].

A very short derivation of this normal form using ideals and group theory was given by VAN DER WAERDEN[5].

Other derivations of the normal form have been given by W. O. MENGE[6], J. H. M. WEDDERBURN[7], O. SCHREIER and E. SPERNER[8], and M. H. INGRAHAM[9].

---

[1] KRULL, W.: Über Begleitmatrizen und Elementarteilertheorie. Freiburg 1921.

[2] DICKSON, L. E.: Modern algebraic theories, Chap. V. Chicago 1926.

[3] BENNETT, A. A.: Amer. Math. Monthly II Vol. 38 (1931) pp. 377–383.

[4] TURNBULL, H. W., and A. C. AITKEN: An introduction to the theory of canonical matrices, Chap. V. London 1932.

[5] VAN DER WAERDEN: Moderne Algebra Vol. II p. 135. Berlin 1931.

[6] MENGE, W. O.: Construction of canonical forms for a linear transformation. Univ. of Michigan dissertation 1931.

[7] WEDDERBURN, J. H. M.: Notes on the theory of matrices. Princeton Univ. 1931.

[8] SCHREIER, O., and E. SPERNER: Vorlesungen über Matrizen, p. 91. Leipzig 1932.

[9] INGRAHAM, M. H.: Abstr. Bull. Amer. Math. Soc. Vol. 38 (1932) p. 814.

The relation between the rational and irrational canonical forms in the case of a matrix having distinct characteristic roots was clearly indicated by I. SCHUR[1]. If A has the distinct characteristic roots $\varrho_1, \varrho_2, \ldots, \varrho_n$ and the characteristic equation

$$\varphi(x) = x^n - c_1 x^{n-1} - c_2 x^{n-2} - \cdots - c_n = 0,$$

and if

$$R = \begin{Vmatrix} c_1 & c_2 & \ldots & c_{n-1} & c_n \\ 1 & 0 & \ldots & 0 & 0 \\ \cdot & \cdot & \cdot & \cdot & \cdot \\ 0 & 0 & \ldots & 1 & 0 \end{Vmatrix}, \qquad P = \begin{Vmatrix} \varrho_1 & 0 & \ldots & 0 \\ 0 & \varrho_2 & \ldots & 0 \\ \cdot & \cdot & \cdot & \cdot \\ 0 & 0 & \ldots & \varrho_n \end{Vmatrix},$$

then $P = Q^1 R Q$ where

$$Q = \begin{Vmatrix} \varrho_1{}^{n-1} & \varrho_2{}^{n-1} & \ldots & \varrho_n{}^{n-1} \\ \varrho_1{}^{n-2} & \varrho_2{}^{n-2} & \ldots & \varrho_n{}^{n-2} \\ \cdot & \cdot & \cdot & \cdot \\ 1 & 1 & \ldots & 1 \end{Vmatrix}.$$

The transforming matrix which corresponds to Q when A is general was given explicitly by TURNBULL and AITKEN[2].

40. WEYR's characteristic. Let A be a matrix with elements in an algebraically closed field $\mathfrak{F}$, and having the characteristic roots $\lambda_1, \lambda_2, \ldots, \lambda_k$. If λ_i is of multiplicity α_i, let

$$\alpha_{i1}, \quad \alpha_{i1} + \alpha_{i2}, \quad \alpha_{i1} + \alpha_{i2} + \alpha_{i3}, \quad \ldots, \quad \alpha_{i1} + \cdots + \alpha_{ip} = \alpha_i$$

be the nullities (§ 8) of the succesive powers

$$A - \lambda_i I, \quad (A - \lambda_i I)^2, \quad \ldots, \quad (A - \lambda_i I)^p$$

respectively, where p is the first integer giving the maximum nullity α_i. The set of numbers $\alpha_{i1}, \ldots, \alpha_{i\varrho}$ is called the WEYR *characteristic* of A relative to λ_i.[3]

If the elementary divisors of A are $(\lambda - \lambda_i)^{e_{il}}$ $(i = 1, \ldots, k)$, then A is similar to a direct sum of matrices

$$J_{il} = \begin{Vmatrix} \lambda_i & 1 & 0 & \ldots & 0 \\ 0 & \lambda_i & 1 & \ldots & 0 \\ \cdot & \cdot & \cdot & \cdot & \cdot \\ 0 & 0 & 0 & \ldots & \lambda_i \end{Vmatrix}$$

(Corollary 39.11) where J_{il} is of order l. Write $J_{il} = \lambda_i I_l + N_l$ where $I_l = (\delta_{rs})$ and $N_l = (\delta_{r+1,s})$. Then

$$I_l^2 = I_l, \qquad I_l N_l = N_l I_l = N_l,$$

$$N_l^2 = (\Sigma_t \delta_{r+1,t} \delta_{t+1,s}) = (\delta_{r+2,s})$$

[1] SCHUR, I.: Trans. Amer. Math. Soc. Vol. 10 (1909) pp. 159—175.

[2] TURNBULL and AITKEN: Canonical matrices, Chap. VI. London 1932.

[3] WEYR, E.: C. R. Acad. Sci., Paris Vol. 100 (1885) pp. 966—969 — Mh. Math. Phys. Vol. 1 (1890) pp. 163—236.

which is of nullity $\nu(N_l^2) = 2$. And in general

$$\nu(N_l^h) = h. \qquad (h = 1, 2, \ldots, l)$$

One may call I_l the *idempotent* and N_l the *nilpotent* matrix corresponding to the elementary divisor $(\lambda - \lambda_i)^{e_{il}}$.[1]

Theorem 40.1. The WEYR *characteristic and the* SEGRE *characteristic* (§ 29) *of A relative to the characteristic root λ_i of multiplicity α_i are conjugate partitions of α_i.*[2]

Suppose the SEGRE characteristic of A relative to λ_i to be written as the rows of a FERRERS diagram[3].

For instance if $(e_{i1}, e_{i2}, e_{i3}, e_{i4}) = (5, 4, 2, 2,)$ we write

$$\begin{matrix} \cdot & \cdot & \cdot & \cdot & \cdot \\ \cdot & \cdot & \cdot & \cdot & \\ \cdot & \cdot & & & \\ \cdot & \cdot & & & \end{matrix}$$

The drop in rank (or increase in nullity) in the successive powers of $A - \lambda_i I$ is evidently the number of dots in the successive columns of the diagram, for every block corresponding to an elementary divisor drops one in rank with successively increasing exponents until 0 is reached, after which no change occurs.

Corollary 40.1. The WEYR *characteristics of a matrix A constitute a complete set of invariants of structure.* (Cf. § 29.)

In other words, two matrices are similar if and only if they have the same characteristic roots and the same WEYR characteristic[4].

Expositions of WEYR's theory have been given by W. H. METZLER[5], K. HENSEL[6], and J. WELLSTEIN[7].

W. O. MENGE[8] proved that if $h_i(\lambda)$ of degree d_i is the i-th invariant factor of $\lambda I - A$, then $h_i(A)$ has the minimum rank of all matrices $f(A)$ where $f(\lambda)$ is a polynomial in λ of degree $\leqq d_i$.

L. AUTONNE[9] proved that a necessary and sufficient condition in order that $AB \overset{S}{=} BA$ is that A^i and B^i have the same ranks for $i = 1, 2, \ldots, n$.

W. KRULL[10] extended WEYR's theory to a general field.

---

1 WEDDERBURN, J. H. M.: Ann. of Math. II Vol. 23 (1921) p. 123.

2 TURNBULL and AITKEN: Canonical matrices, p. 80. London 1932.

3 MACMAHON, P. A.: Combinatory analysis Vol. II p. 3. Cambridge Univ. Press 1915.

4 WEYR, E.: l. c.

5 METZLER, W. H.: Amer. J. Math. Vol. 14 (1892) pp. 326—377.

6 HENSEL, K.: J. reine angew. Math. Vol. 127 (1904) pp. 116—166.

7 WELLSTEIN, J.: J. reine angew. Math. Vol. 163 (1930) pp. 166—182.

8 MENGE, W. O.: Bull. Amer. Math. Soc. Vol. 38 (1932) pp. 88—94.

9 AUTONNE, L.: Nouv. Ann. Math. IV Vol. 12 (1912) pp. 118—127.

10 KRULL, W.: Über Begleitmatrizen und Elementarteilertheorie. Freiburg 1921.

41. Unitary and orthogonal equivalence. If there exists a unitary matrix (§18) U such that

$$A = U^{CT} B U,$$

then A is both conjunctive with and similar to B. We shall write $A \overset{U}{=} B$ to mean that *A is conjunctive with B by a unitary transformation.* If all the matrices are real so that U is an orthogonal matrix, we say that *A is congruent with B by an orthogonal transformation,* and write $A \overset{O}{=} B$. Many properties of the relation $\overset{O}{=}$ are implied by the corresponding properties of the relation $\overset{U}{=}$.

Theorem 41.1. If A and B are hermitian in the complex field $\mathfrak{C}$, a necessary and sufficient condition in order that $A \overset{U}{=} B$ is that $\lambda I - A \overset{E}{=} \lambda I - B$.

By Theorem 36.5 the condition $\lambda I - A \overset{E}{=} \lambda I - B$ implies the existence of a non-singular matrix P such that

$$P^{CT} I P = I, \qquad P^{CT} B P = A.$$

The first equation indicates that P is unitary. Then, from the second, $A \overset{U}{=} B$.

Conversely, if $A \overset{U}{=} B$, there exists a unitary matrix U $(U^{CT} I U = I)$ such that $U^{CT} B U = A$. Then by Theorem 36.5, $\lambda I - A \overset{E}{=} \lambda I - B$.

Corollary 41.1. If A and B are orthogonal in the real field $\mathfrak{R}$, a necessary and sufficient condition in order that $A \overset{O}{=} B$ is that $\lambda I - A \overset{E}{=} \lambda I - B$.

Theorem 41.2. If H is hermitian,

$$H \overset{U}{=} [\lambda_1, \lambda_2, \ldots, \lambda_n],$$

where $\lambda_1, \lambda_2, \ldots, \lambda_n$ are the characteristic roots of H.

By Theorem 36.8 the elementary divisors of H are simple, so the cogredient normal form is the diagonal matrix $[\lambda_1, \lambda_2, \ldots, \lambda_n]$.

Direct proofs of this theorem for H real and symmetric and U orthogonal were given by L. STICKELBURGER[1] and, using infinitesimal transformations, by J. J. SYLVESTER[2].

Theorem 41.3. If A is any matrix with complex elements, then $A \overset{U}{=} B$, where[3]

$$B = \begin{Vmatrix} \lambda_1 & 0 & 0 & \ldots & 0 \\ b_{21} & \lambda_2 & 0 & \ldots & 0 \\ b_{31} & b_{32} & \lambda_3 & \ldots & 0 \\ \cdot & \cdot & \cdot & \cdot & \cdot \\ b_{n1} & b_{n2} & b_{n3} & \ldots & \lambda_n \end{Vmatrix}.$$

Let λ_1 be a characteristic root of A, and $(q_1, q_2, \ldots, q_n)$ the corresponding pole, so that

$$\sum_{i=1}^{n} a_{ij} q_i = \lambda_1 q_j. \qquad (j = 1, 2, \ldots, n)$$

[1] STICKELBURGER, L.: Progr. der eidgen. polyt. Schule. Zürich 1877.
[2] SYLVESTER, J. J.: Mess. of Math. Vol. 19 (1890) pp. 1–5.
[3] SCHUR, I.: Math. Ann. Vol. 66 (1909) pp. 488–510.

The sum $\sum \bar{q}_j q_j = q^2$ is positive. Set $q_{j1} = q_j/q$. Then

$$\sum a_{ij} q_{i1} = \lambda_1 q_{j1}, \qquad \sum \bar{q}_{i1} q_{i1} = 1 .$$

It is possible to determine $q_{i2}, \ldots, q_{in}$ so that $Q = (q_{rs})$ is unitary. Then $Q^{\mathrm{T}} A Q^{\mathrm{C}}$ is of the form

$$\begin{Vmatrix} \lambda_1 & 0 & 0 & \ldots & 0 \\ b_{21} & b_{22} & b_{23} & \ldots & b_{2n} \\ \cdot & \cdot & \cdot & \cdot & \cdot \end{Vmatrix}.$$

Continue with the last $n-1$ rows and columns. Since the product of unitary matrices is unitary, we have the theorem.

Theorem 41.1 also appears as a corollary to this theorem. For if A is hermitian, so is B, so that B is diagonal.

A matrix A is called *normal* if $A^{\mathrm{CT}} A = A A^{\mathrm{CT}}$.[1]

Evidently A is normal if A^{CT} is equal to a rational function of A, so normal matrices include hermitian, skew hermitian, unitary, orthogonal, symmetric and skew matrices as special instances.

Theorem 41.4. A necessary and sufficient condition in order that $A \overset{\mathrm{U}}{=} D$ where D is diagonal is that A be normal[2].

The property of being normal is a unitary invariant, for if $A = U^{\mathrm{CT}} B U$, then

$$A A^{\mathrm{CT}} = U^{\mathrm{CT}} B B^{\mathrm{CT}} U, \qquad A^{\mathrm{CT}} A = U^{\mathrm{CT}} B^{\mathrm{CT}} B U.$$

If $B B^{\mathrm{CT}} = B^{\mathrm{CT}} B$, then $A A^{\mathrm{CT}} = A^{\mathrm{CT}} A$.

By Theorem 41.3 we may take

$$B = \begin{Vmatrix} \lambda_1 & 0 & 0 & \ldots \\ b_{21} & \lambda_2 & 0 & \ldots \\ b_{31} & b_{32} & \lambda_3 & \ldots \\ \cdot & \cdot & \cdot & \cdot \end{Vmatrix}.$$

The element in the $(1, 1)$-position in $B B^{\mathrm{CT}}$ is λ_1^2, the corresponding element in $B^{\mathrm{CT}} B$ is $\lambda_1^2 + \bar{b}_{21} b_{21} + \bar{b}_{31} b_{31} + \cdots$ Hence $b_{21} = b_{31} = \cdots = 0$. Comparison of elements in the $(2, 2)$-position now shows that $b_{32} = b_{42} = \cdots = 0$, etc. Hence if A is normal, it can be reduced to diagonal form. Since two diagonal matrices are commutative, the converse is obvious.

Theorem 41.4. If A is positive semi-definite hermitian, and m is any positive integer, there exists a unique positive semi-definite hermitian matrix B such that $B^m = A$. It is of the same rank as A.[3]

By Theorem 41.4 there exists a unitary matrix U such that

$$U^{\mathrm{CT}} A U = [\lambda_1, \lambda_2, \ldots, \lambda_n] = L, \qquad \lambda_i \geqq 0 .$$

[1] Toeplitz, O.: Math. Z. Vol. 2 (1918) pp. 187—197.

[2] Toeplitz, O.: l. c.

[3] Autonne: Rend. Circ. mat. Palermo Vol. 16 (1902) pp. 104—128 — Bull. Soc. Math. France Vol. 31 (1903) pp. 140—155.

Let
$$M = [\mu_1, \mu_2, \ldots, \mu_n], \qquad \mu_i^m = \lambda_i, \qquad \mu_i \geqq 0.$$
Set $P = UMU^{CT}$. Then $P^m = A$.

Suppose that Q is another semi-definite hermitian matrix such that $Q^m = A$. Then $N^m = L$ where $N = UQU^{CT}$. By Theorem 16.1 the characteristic roots of L are the m-th powers of the roots of N, and since the latter are all $\geqq 0$, they are the μ's. Hence by Theorem 41.2 there exists a unitary matrix V such that
$$N = V^{CT} M V.$$
Since $VN^m = M^m V$, V is commutative with L. Suppose that the λ's are grouped into sets of distinct roots so that
$$\lambda_1 = \lambda_2 = \cdots = \lambda_\alpha, \qquad \lambda_{\alpha+1} = \cdots = \lambda_{\alpha+\beta}, \qquad \ldots.$$
It follows from $VL = LV$ that $v_{ij} = 0$ if $\lambda_i \neq \lambda_j$, so that
$$V = V_1 \dotplus V_2 \dotplus \cdots \dotplus V_k,$$
where V_1 is of order α, V_2 of order β, etc. Evidently each V_i is unitary. Then
$$N = V_1^{CT} \mu_\alpha I_\alpha V_1 \dotplus V_2^{CT} \mu_{\alpha+\beta} I_\beta V_2 \dotplus \cdots = M.$$
Hence $Q = U^{CT} N U = U^{CT} M U = P$, and P is unique.

Corollary 41.4. If A is positive semi-definite hermitian, there exists a unique positive semi-definite hermitian matrix P of the same rank as A such that $A = P^{CT} P$.[1]

Theorem 41.5. Every non-singular matrix is uniquely expressible as a product of a unitary matrix by a positive definite hermitian matrix[2].

For $A^{CT} A$ is positive definite hermitian and hence by Corollary 41.4 equals $B^{CT} B$, where B is positive definite hermitian. Then $AB^{I} = U$ is unitary, since $U^{CT} U = I$, and $A = UB$. If $A = VC$ where V is unitary, then $A^{CT} A = C^{CT} C$. If C is positive definite hermitian, $C = B$ by Corollary 41.4.

Every normal matrix can be represented as a product of a semi-definite symmetric matrix P and an orthogonal matrix, and conversely, every such product is normal[3].

R. Weitzenböck[4] gave two methods for solving the equation $XX^T = A$ for A symmetric and non-singular.

---

[1] Autonne, L.: Bull. Soc. Math. France Vol. 31 (1903) pp. 140–155.

[2] Autonne, L.: Bull. Soc. Math. France Vol. 30 (1902) pp. 121–134. — Wintner, A., and F. D. Murnaghan: Proc. Nat. Acad. Sci. U.S.A. Vol. 17 (1931) pp. 676–678.

[3] Murnaghan, F. D., and A. Wintner: Proc. Nat. Acad. Sci. U.S.A. Vol. 17 (1931) pp. 417–420.

[4] Weitzenböck, R.: Akad. Wetensch. Amsterdam, Proc. Vol. 35 (1932) pp. 328–330.

Theorem 41.6. If A is non-singular, there exist two unitary matrices U and V such that

$$UAV = [\mu_1, \mu_2, \ldots, \mu_n],$$

where $\mu_1, \mu_2, \ldots, \mu_n$ are the positive square roots of the characteristic roots of AA^{CT}.[1]

Since $A^{CT}A$ is positive semi-definite hermitian, there exists a unitary matrix V such that

$$V^{CT}A^{CT}AV = L = [\lambda_1, \lambda_2, \ldots, \lambda_n], \qquad \lambda_i \geqq 0.$$

Set $AV = K$ so that $K^{CT}K = L$. Set

$$\mu_i = +\sqrt{\lambda_i}, \qquad k_{rs} = w_{rs}\mu_s, \qquad K = WM,$$

where $M = [\mu_1, \mu_2, \cdots, \mu_n]$. Then

$$\sum \bar{w}_{ir} w_{is} = \sum k_{ir} k_{is}/\mu_r \mu_s = \delta_{rs},$$

so that W is unitary. Let $U = W^I$. Then $UAV = M$.

This theorem was first proved for the real case, U and V being orthogonal, by E. Beltrami[2] and C. Jordan[3]. It was treated again by J. J. Sylvester[4].

Corollary 41.6. If U and V are unitary matrices such that $UAV = M$ is diagonal and real, then

$$V^{CT}A^{CT}AV = M^2 = UAA^{CT}U^{CT}.$$[5]

For

$$M = M^{CT} = V^{CT}A^{CT}U^{CT},$$

and

$$M^2 = V^{CT}A^{CT}U^{CT}UAV = V^{CT}A^{CT}AV.$$

Schläfli[6] discussed the reduction of an orthogonal matrix by a similarity transformation.

H. Hilton[7] showed that an orthogonal matrix can be transformed by orthogonal matrices into a direct sum of orthogonal matrices each of which has a characteristic matrix with elementary divisors (1) $(\lambda - a)^r$ and $(\lambda - 1/a)^r$, $a \neq 0$, or (2) $(\lambda - 1)^r$ or $(\lambda + 1)^r$, r odd, or (3) $(\lambda + 1)^r$ and $(\lambda + 1)^r$, r even.

42. The structure of unitary and orthogonal matrices. A unitary matrix is both a conjunctive automorph and a similarity automorph of the identity matrix I. An orthogonal matrix is a real unitary matrix.

---

1 Autonne, L.: Ann. Univ. Lyon II Vol. 38 (1915) pp. 1—77.

2 Beltrami, E.: Giorn. Mat. Battaglini Vol. 11 (1873) pp. 98—106.

3 Jordan, C.: J. Math. pures appl. II Vol. 19 (1874) pp. 35—54.

4 Sylvester, J. J.: C. R. Acad. Sci., Paris Vol. 108 (1889) pp. 651—653 — Mess. of Math. Vol. 19 (1890) pp. 42—46.

5 Cosserat, E.: Ann. Fac. Sci. Univ. Toulouse Vol. 3 (1889) M. 1—12.

6 Schläfli: J. reine angew. Math. Vol. 65 (1866) pp. 185—187.

7 Hilton, H.: Mess. of Math. Vol. 41 (1912) pp. 146—154.

The literature of orthogonal matrices and their determinants is extensive. Rather complete bibliographies are given by E. PASCAL[1] and T. MUIR[2].

Theorem 42.1. If K is a skew-hermitian matrix, then

$$U = (I + K)^{\mathrm{I}} (I - K)$$

is unitary. There are no other unitary matrices U having $d(I + U) \neq 0$.[3]

This is a special case of Theorem 37.1. The hypothesis of the latter that $d(I + K) \neq 0$ and $d(I - K) \neq 0$ is automatically fulfilled, since (Corollary 18.32) the skew-hermitian K has only purely imaginary characteristic roots.

The exceptional case $d(I + U) = 0$ makes considerable difficulty, and as the characteristic roots of a unitary matrix have absolute value unity, it is not very exceptional. Since CAYLEY[4] first proved the above theorem for orthogonal matrices of order 4 by long calculation, many ways of avoiding the difficulty have been suggested[5].

L. LOEWY[6] showed that every unitary matrix is of the form

$$U = \theta (I + K)^{\mathrm{I}} (I - K),$$

where K is skew-hermitian and θ is a root of unity.

The representation of an orthogonal matrix as a product of orthogonal matrices of simple type was discussed geometrically by A. VOSS[7] and E. GOURSAT[8], and algebraically by L. KRONECKER[9].

Defining an inversion to be a real orthogonal matrix S such that $d(S) = -1$, $d(\lambda I - S) = (\lambda - 1)^{n-1} (\lambda + 1)$, L. AUTONNE[10] proved that every real orthogonal matrix is a product of inversions.

G. VITALI[11] showed that a real orthogonal matrix is a product of rotations in a space of $n - 2$ dimensions, and a reflection if its determinant is -1.

---

1 PASCAL, E.: Die Determinanten, pp. 157–175. Teubner 1900.

2 MUIR, T.: Proc. Roy. Soc. Edinburgh Vol. 47 (1926–1927) pp. 252–282.

3 LOEWY, A.: C. R. Acad. Sci., Paris Vol. 123 (1896) pp. 168–171. — AUTONNE, L.: Rend. Circ. mat. Palermo Vol. 16 (1902) pp. 104–128.

4 CAYLEY: J. reine angew. Math. Vol. 32 (1846) pp. 119–123.

5 METZLER, W. H.: Amer. J. Math. Vol. 15 (1892) pp. 274–282. — PRYM, F.: Abh. Ges. Wiss. Göttingen Vol. 38 (1892) pp. 1–42. — TABER, H.: Proc. London Math. Soc. Vol. 24 (1892) pp. 290–306 — Proc. Amer. Acad. Arts Sci. Vol. 28 (1892–1893) pp. 212–221 — Amer. J. Math. Vol. 16 (1893) pp. 123–130.

6 LOEWY, L.: Nova Acta. Abh. der Kaiserl. Leop.-Carol. Akad. Vol. 71 No. 8 pp. 379–446 — Nachr. Ges. Wiss. Göttingen Vol. 3 (1900) pp. 298–303.

7 VOSS, A.: Math. Ann. Vol. 13 (1878) pp. 320–374.

8 GOURSAT, E.: Ann. École norm. III Vol. 6 (1889) pp. 1–102.

9 KRONECKER, L.: S.-B. preuß. Akad. Wiss. 1890 pp. 525–541, 601–607, 691–699, 873–885, and 1063–1080.

10 AUTONNE, L.: C. R. Acad. Sci., Paris Vol. 136 (1903) pp. 1185–1186 — Ann. Univ. Lyon II Vol. 12 (1903) pp. 1–124.

11 VITALI, G.: Boll. Un. Mat. Ital. Vol. 7 (1928) pp. 1–7.

T. J. STIELTJES[1] showed that if A and B are third order orthogonal matrices of determinant $+1$, then $A+B$ is never of rank 2. E. NETTO[2] extended this to show that if $A=(a_{rs})$ and $B=(b_{rs})$ are orthogonal, ε_r can be chosen $+1$ or -1 so that $d(a_{rs}+\varepsilon_r b_{rs})$ vanishes for n even only if every element is 0, and for n odd only if all minors of order 2 vanish.

A. LOEWY[3] showed that STIELTJES' theorem is a consequence of the following theorem of FROBENIUS[4]: The elementary divisors of the characteristic matrix of an orthogonal matrix vanish for reciprocal values except that $\lambda-1$ and $\lambda+1$ may occur with odd exponents.

L. TOSCANO[5] established for involutory matrices a theorem analogous to that of STIELTJES for orthogonal matrices.

A. VOSS[6] proved that every non-singular matrix may be represented as a product of two symmetric matrices in infinitely many ways. FROBENIUS[7] proved that every matrix may be represented as a product of two symmetric matrices, one of which is non-singular.

H. HILTON[8] proved the above theorem of FROBENIUS, and also proved that a matrix can be represented as a product of two skew matrices if and only if the elementary divisors of its characteristic matrix occur in pairs $(\lambda-a)^r$, $(\lambda-a)^r$; and as a product of a symmetric matrix and a skew matrix if and only if the elementary divisors occur in pairs $(\lambda-a)^r$, $(\lambda+a)^r$. This same conclusion was reached by H. STENZEL[9].

O. TOEPLITZ[10] proved that, if H is hermitian, there exists a matrix P with 0's above the main diagonal such that $H=PP^{CT}$.

E. SCHMIDT[11] showed that for every A there is a K with 0's above the main diagonal such that KA is unitary. This is equivalent to TOEPLITZ' theorem.

FROBENIUS[12] defined the *span* (Spannung) of A to be $s(A)=t(AA^{CT})$. If U is unitary, $s(UA)=s(A)=s(AU)$, and $s(A-U)=s(I-U^{I}A)$.

J. RADON[13] determined the maximum number $\nu(n)$ of matrices such that $x_1A_1+\cdots+x_\nu A_\nu$ is orthogonal for all x's. If $n=2^{4\alpha+\beta}n'$, n' odd, $\beta=0,1,2,3$, this value is $\nu=2^\beta+8\alpha$.

[1] STIELTJES, T. J.: Acta math. Vol. 6 (1885) pp. 319–320.
[2] NETTO, E.: Acta math. Vol. 9 (1887) pp. 295–300.
[3] LOEWY, A.: E. Pascal's Repertorium der höheren Mathematik Vol. I Chap. II. Teubner 1910.
[4] FROBENIUS: J. reine angew. Math. Vol. 84 (1878) p. 48.
[5] TOSCANO, L.: Rend. Roy. Inst. Lombardo IIa Vol. 61 (1928) pp. 187–195.
[6] VOSS, A.: Math. Ann. Vol. 13 (1878) pp. 320–374.
[7] FROBENIUS: S.-B. preuß. Akad. Wiss. 1910 pp. 3–15.
[8] HILTON, H.: Homogeneous linear substitutions. Oxford 1914.
[9] STENZEL, H.: Math. Z. Vol. 15 (1922) pp. 1–25.
[10] TOEPLITZ, O.: Nachr. Ges. Wiss. Göttingen 1907 pp. 101–109.
[11] SCHMIDT, E.: Math. Ann. Vol. 63 (1907) pp. 433–476.
[12] FROBENIUS: S.-B. preuß. Akad. Wiss. 1911 pp. 241–248.
[13] RADON, J.: Abh. math. Semin. Hamburg. Univ. Vol. 1 (1921) pp. 1–14.

F. D. MURNAGHAN[1] found that a matrix of order n has $2n+(n-1)^2$ independent functions of its elements which are invariant under unitary transformation.

VII. Composition of matrices.

43. Direct sum and direct product. The present chapter is concerned with the set of all matrices of finite order with elements in a ring or field. The operations of addition and multiplication can be applied only to matrices of the same order, but other operations will be defined according to which matrices of different orders may be combined. Also *invariant matric functions* $T(A)$ will be defined, where $T(A)$ is not necessarily of the same order as A, which have the property that $T(AB) = T(A)\,T(B)$.

A brief treatment of this topic with many references to original sources is given by A. LOEWY in PASCAL's *Repertorium der höheren Mathematik* Vol. I pp. 138—153. Teubner 1910.

Let $A = (a_{rs})$ of order α and $B = (b_{rs})$ of order β be two matrices with elements in a ring $\mathfrak{R}$. The *direct sum* (cf. § 29) of A and B is defined to be

$$A \dotplus B = \begin{Vmatrix} A & 0 \\ 0 & B \end{Vmatrix}.$$

It is of order $\alpha+\beta$.

Theorem 43.1. The following identities hold:

(a) $(A \dotplus B) \dotplus C = A \dotplus (B \dotplus C)$,

(b) $(A_1 + A_2) \dotplus (B_1 + B_2) = (A_1 \dotplus B_1) + (A_2 \dotplus B_2)$,

(c) $(A_1 \dotplus B_1)(A_2 \dotplus B_2) = A_1 A_2 \dotplus B_1 B_2$,

(d) $(A \dotplus B)^{\mathrm{T}} = A^{\mathrm{T}} \dotplus B^{\mathrm{T}}$,

(e) $(A \dotplus B)^{\mathrm{I}} = A^{\mathrm{I}} \dotplus B^{\mathrm{I}}$,

(f) $d(A \dotplus B) = d(A)\,d(B)$,

(g) $t(A \dotplus B) = t(A) + t(B)$.

The proofs follow directly from the definition of direct sum. Of course (e) holds only when A^{I} and B^{I} exist.

The *left direct product* is defined as

$$A \cdot\!\times B = \begin{Vmatrix} A b_{11} & \dots & A b_{1\beta} \\ \dots & \dots & \dots \\ A b_{\beta 1} & \dots & A b_{\beta\beta} \end{Vmatrix}.$$

The *right direct product* $A \times\!\cdot B$ is defined similarly and has similar properties. The order of each direct product is $\alpha\beta$. Evidently $A \cdot\!\times B = B \times\!\cdot A$ if A and B have elements in a commutative ring.

[1] MURNAGHAN, F. D.: Proc. Nat. Acad. Sci. U.S.A. Vol. 18 (1932) pp. 185—189.

The concept of direct product of matrices arises naturally from the concept of direct product in group theory[1].

Theorem 43.2. The following identities hold:

(a) $(A \cdot\!\times B) \cdot\!\times C = A \cdot\!\times (B \cdot\!\times C)$,

(b) $(A_1 + A_2) \cdot\!\times B = A_1 \cdot\!\times B + A_2 \cdot\!\times B$,

(c) $(A \cdot\!\times B)^{\mathrm{T}} = A^{\mathrm{T}} \cdot\!\times B^{\mathrm{T}}$,

(d) $t(A \cdot\!\times B) = t(A)\, t(B)$.

These are all evident from the definition of direct product.

Theorem 43.3. If A is of order α and B is of order β, and if $A \cdot\!\times B = C = (c_{rs})$, then

$$c_{rs} = a_{r_1 s_1} b_{r_2 s_2},$$

where

$$r - 1 = \alpha(r_2 - 1) + r_1 - 1, \quad 0 \leqq r_1 - 1 < \alpha,$$
$$s - 1 = \alpha(s_2 - 1) + s_1 - 1, \quad 0 \leqq s_1 - 1 < \alpha.$$

This also follows directly from the definition.

Theorem 43.4. If $\mathfrak{R}$ is commutative[2],

$$A_1 A_2 \cdot\!\times B_1 B_2 = (A_1 \cdot\!\times B_1)(A_2 \cdot\!\times B_2).$$

For by Theorem 43.3

$$\begin{aligned}(A_1 \cdot\!\times B_1)(A_2 \cdot\!\times B_1) &= (a_{1 r_1 s_1} b_{1 r_2 s_2})(a_{2 r_1 s_1} b_{2 r_2 s_2}) \\ &= (\Sigma_h a_{1 r_1 h_1} b_{1 r_2 h_2} a_{2 h_1 s_1} b_{2 h_2 s_2}) \\ &= (\Sigma_{h_1} a_{1 r_1 h_1} a_{2 h_1 s_2} \Sigma_{h_2} b_{1 r_2 h_2} b_{2 h_2 s_2}) \\ &= A_1 A_2 \cdot\!\times B_1 B_2,\end{aligned}$$

where it is understood that

$$h - 1 = \alpha(h_2 - 1) + h_1 - 1, \quad 0 \leqq h_1 - 1 < \alpha.$$

Corollary 43.4. If A of order α and B of order β have elements in a field, then

(a) $A \cdot\!\times B = (A \cdot\!\times I_\beta)(I_\alpha \cdot\!\times B)$,

(b) $d(A \cdot\!\times B) = d(A)^\beta d(B)^\alpha$,

(c) $(A \cdot\!\times B)^{\mathrm{I}} = A^{\mathrm{I}} \cdot\!\times B^{\mathrm{I}}$,

where I_α is the identity matrix of order α.

Property (b) was proved by K. HENSEL[3], E. NETTO[4], and R. D. VON STERNECK[5].

---

[1] HÖLDER, O.: Math. Ann. Vol. 43 (1893) p. 305. — BURNSIDE, W.: Theory of groups of finite order, p. 40. Cambridge 1911.

[2] STÉPHANOS, C.: J. Math. pures appl. V Vol. 6 (1900) pp. 73—128. — AITKEN, A. C.: Proc. Edinburgh Math. Soc. II Vol. 1 (1927) pp. 135—138.

[3] HENSEL, K.: Acta math. Vol. 14 (1891) pp. 317—319.

[4] NETTO, E.: Acta math. Vol. 17 (1893) pp. 199—204.

[5] STERNECK, R. D. VON: Mh. Math. Phys. Vol. 6 (1895) pp. 205—207.

Property (c), originally given by STÉPHANOS[1], was proved in tensor notation by L. H. RICE[2].

Theorem 43.5. If $PAP_1 = A_1$, *and* $QBQ_1 = B_1$, *where each matrix has elements in a commutative ring, then*

(a) $(P \dot{+} Q)(A \dot{+} B)(P_1 \dot{+} Q_1) = A_1 \dot{+} B_1$.
(b) $(P \cdot\!\times Q)(A \cdot\!\times B)(P_1 \cdot\!\times Q_1) = A_1 \cdot\!\times B_1$.

The proof of (a) follows from Theorem 43.1 c, and that of (b) follows from Theorem 43.4.

Corollary 43.5. In a principal ideal ring

$$\varrho(A \dot{+} B) = \varrho(A) + \varrho(B),$$
$$\varrho(A \cdot\!\times B) = \varrho(A)\,\varrho(B),$$

where $\varrho(A)$ *denotes the rank of* A.

For, by Theorem 26.2, A_1 and B_1 may be chosen in the form $[h_1, \ldots, h_\varrho, 0, \ldots, 0]$.

Theorem 43.6. If A *and* B *are symmetric with elements in an ordered field with characteristic* $\neq 2$,

$$\sigma(A \dot{+} B) = \sigma(A) + \sigma(B),$$
$$\sigma(A \cdot\!\times B) = \sigma(A)\,\sigma(B),$$

where $\sigma(A)$ *denotes the signature of* A.

For it may be assumed that A and B are in diagonal form with p_1 and p_2 positive elements, and n_1 and n_2 negative elements, respectively. Then

$$\sigma(A \dot{+} B) = p_1 + p_2 - n_1 - n_2 = \sigma(A) + \sigma(B).$$

Since $A \cdot\!\times B$ is diagonal, and since its diagonal elements are the $\alpha\beta$ products of the diagonal elements of A by those of B,

$$\sigma(A \cdot\!\times B) = (p_1 p_2 + n_1 n_2) - (p_1 n_2 + p_2 n_1)$$
$$= (p_1 - n_1)(p_2 - n_2) = \sigma(A)\,\sigma(B).$$

In the rest of this section we shall understand that all matrices have elements in an algebraically closed field $\mathfrak{C}$.

Theorem 43.7. The characteristic roots of $A \dot{+} B$ *are the characteristic roots of* A *together with those of* B.

This is obvious from the definition of direct sum.

Let $\varphi(\xi, \eta) = \sum c_{ij}\, \xi^i \eta^j$ be a polynomial in ξ and η, and define

$$\varphi(A; B) = \sum c_{ij}\, A^i \cdot\!\times B^j.$$

Theorem 43.8. The characteristic roots of $\varphi(A; B)$ *are the* $\alpha\beta$ *functions* $\varphi(a_p, b_q)$ *where the a's are the characteristic roots of* A *and the b's are those of* B.[3]

[1] STÉPHANOS: l. c.

[2] RICE, L. H.: J. Math. Physics, Massachusetts Inst. Technol. Vol. 5 (1925) pp. 55–64.

[3] STÉPHANOS, C.: J. Math. pures appl. V Vol. 6 (1900) pp. 73–128.

Determine P and Q so that $PAP^{\mathrm{I}} = A_1$ and $QBQ^{\mathrm{I}} = B_1$ have 0's above the main diagonal (Corollary 39.1) and their characteristic roots in the main diagonal. This form is maintained under addition, multiplication, scalar multiplication, and direct multiplication. If the characteristic roots of A_1 are $a_1, \ldots, a_\alpha$ and those of B_1 are $b_1, \ldots, b_\beta$, those of $\sum c_{ij} A^i \cdot\!\times B^j$ are $\sum c_{ij} a_p{}^i b_p{}^j$. Since

$$A_1 \cdot\!\times B_1 = (P \cdot\!\times Q)(A \cdot\!\times B)(P \cdot\!\times Q)^{\mathrm{I}},$$

the characteristic roots of $A \cdot\!\times B$ are the same as those of $A_1 \cdot\!\times B_1$.

Corollary 43.81. The characteristic roots of $A \cdot\!\times B$ are the $\alpha\beta$ products $a_i b_j$ of a characteristic root of A by one of B.

The above corollary was applied by STÉPHANOS[1] to find the equation whose $\alpha\beta$ roots are the $\alpha\beta$ values of the function $\varphi(x, y)$ where x and y are determined by

$$f(x) = x^\alpha + a_1 x^{\alpha-1} + \cdots + a_\alpha = 0,$$
$$g(y) = y^\beta + b_1 y^{\beta-1} + \cdots + b_\beta = 0.$$

Let A have the characteristic equation $f(x) = 0$, and B the characteristic equation $g(y) = 0$. The solution of the problem is the equation

$$|\varphi(A; B) - \lambda I_{\alpha\beta}| = 0.$$

This theorem yields a proof of the theorem that every polynomial with rational coefficients in two algebraic numbers is itself a algebraic number. If φ has rational integral coefficients, and if x and y are integral algebraic numbers (so that A and B have rational integral elements), then $\varphi(x, y)$ is integral.

Corollary 43.82. The resultant of $f(x) = 0$ and $g(x) = 0$ is

$$|A \cdot\!\times I_\beta - I_\alpha \cdot\!\times B| = 0,$$

where $f(x) = 0$ is the characteristic equation of A and $g(x) = 0$ is the characteristic equation of B.

This is the special case of Theorem 43.8 where $\varphi(\xi, \eta) = \xi - \eta$.

Another method of composing two matrices was given by SYLVESTER[2] and STÉPHANOS[3], and called by the latter *bialternate composition.* Let A and B be each of order n, and let (r_1, r_2) be the r-th pair of integers in the sequence

$$(1, 2), (1, 3), \ldots, (1, n), (2, 3), \ldots, (2, n), (n-2, n-1), \ldots, (n-1, n)$$

and (s_1, s_2) the s-th pair. Then by definition the bialternate product of A and B is

$$A \cdot B = C = (c_{rs})$$

where

$$c_{rs} = \tfrac{1}{2}(a_{r_1 s_1} b_{r_2 s_2} - a_{r_1 s_2} b_{r_2 s_1} - a_{r_2 s_1} b_{r_1 s_2} + a_{r_2 s_2} b_{r_1 s_1}).$$

[1] STÉPHANOS: l. c.
[2] SYLVESTER: J. reine angew. Math. Vol. 88 (1880) pp. 49–67.
[3] STÉPHANOS: l. c.

It was shown by STÉPHANOS that

$$A \cdot B = B \cdot A, \qquad (A_1 + A_2) \cdot B = A_1 \cdot B + A_2 \cdot B,$$

and, if $A \cdot A$ be written $A^{\dot{2}}$, that

$$(A + B)^{\dot{2}} = A^{\dot{2}} + 2A \cdot B + B^{\dot{2}}.$$

Bialternate composition cannot be associative, for $A \cdot B$ is of order $\frac{1}{2}n(n-1)$, and composition is defined only for matrices of the same order.

44. Product-matrices and power-matrices. If A and B, of orders α and β respectively, are the matrices of the transformations

$$x_i' = \sum a_{ij} x_j, \qquad y_i' = \sum b_{ij} y_j,$$

and if the x's and y's are independent variables, the products

$$x_1 y_1, x_2 y_1, \ldots, x_\alpha y_1, x_1 y_2, \ldots, x_\alpha y_2, \ldots, x_\alpha y_\beta$$

are the variables of a transformation whose matrix is $A \cdot\!\times B$, called the *product transformation* of the two given transformations.

By Theorem 43.2a,

$$D_m(A) = A \cdot\!\times A \cdot\!\times \ldots \cdot\!\times A$$

to m factors is well-defined, and is called the m-th *product-matrix* of A.

Theorem 44.1.

$$D_m(AB) = D_m(A)\, D_m(B).$$

This follows from Theorem 43.4.

Now let the two transformations be identical, and consider the $\frac{1}{2}\alpha(\alpha+1)$ distinct products

$$x_1^2,\ x_1 x_2,\ \ldots,\ x_1 x^\alpha,\ x_2^2,\ x_2 x_3,\ \ldots,\ x_\alpha^2.$$

They define a transformation whose matrix $P_2(A)$ is called the *second power-matrix* of A. That is, $x_1'^2, x_1' x_2', \ldots$ are linear homogeneous functions of $x_1^2, x_1 x_2, \ldots$ whose coefficients are polynomials in the elements of A. Such a system $x_1^2, \ldots$ of polynomials was called a *transformable system* by J. DERUYTS[1]. The m-th power-matrix $P_m(A)$ is similarly defined.

Theorem 44.2.

$$P_m(AB) = P_m(A)\, P_m(B).$$

Consider the corresponding transformable system

$$X_{i_1, \ldots, i_\alpha} = x_1^{i_1} \ldots x_\alpha^{i_\alpha} \qquad i_1 + \cdots + i_\alpha = m.$$

Let A be the matrix by which the x'''s are given in terms of the x''s, and B the matrix by which the x''s are given in terms of the x's. Then AB is the matrix by which the x'''s are given in terms of the x's. The matrix by which the X'''s are given in terms of the X''s is $P_m(A)$,

[1] DERUYTS, J.: Bull. Acad. Sci. Belgique III Vol. 32 (1896) p. 82.

and the matrix by which the X'''s are given in terms of the X's is $P_m(B)$. The matrix by which the X''''s are given in terms of the X's may be obtained as $P_m(A)P_m(B)$ by eliminating the X'''s from the induced transformations, or as $P_m(AB)$ by calculating the transformation induced on the X's by the product transformation AB.

Theorem 44.3. If A is of order α,

$$d\,P_m(A) = d(A)^k, \qquad k = \binom{m+\alpha-1}{\alpha}.$$

Let the elements of A be indeterminates. The power transformation has an inverse if and only if $d(A) \neq 0$. Hence $dP_m(A)$ vanishes when and only when $d(A) = 0$, and since $d(A)$ is irreducible (Theorem 10.1), $dP_m(A) = c\,d(A)^k$, where $c \neq 0$. When $A = I$, the power transformation becomes the identity, so $c = 1$. If A is of order α, then $P_m(A)$ is of order $\binom{\alpha+m-1}{m}$, and each element is of degree m in the elements of A. Hence k is the product of this order by m/α.

This theorem was given for $\alpha = 4$ and $m = 2$ by E. HUNYADY[1]. The fact that $dP_m(A)$ is some power of $d(A)$ was proved by B. IGEL[2]. The theorem as stated was proved by G. VON ESCHERICH[3].

Theorem 44.4. The characteristic roots of $P_m(A)$ are the $\binom{\alpha+m-1}{\alpha}$ products of the m-th degree of the characteristic roots of A.[4]

By Theorem 41.3 it is possible to determine a unitary matrix U such that $U^{CT}AU = B$ has 0's above the main diagonal and its characteristic roots in the main diagonal. By Theorem 44.2,

$$P_m(U^{CT})\,P_m(A)\,P_m(U) = P_m(B).$$

Since the elements $x_1^{i_1}\ldots x_\alpha^{i_\alpha}$ of the transformable system corresponding to $P_m(B)$ are arranged in the lexacographic order, $P_m(B)$ also has 0's above the main diagonal. If the conjugate-transpose of the transformation on the x's is equal to the inverse, the same is true of the transformation induced on the products $x_1^{i_1}\ldots x_\alpha^{i_\alpha}$. Hence

$$P_m(U^{CT}) = P_m(U)^{CT} = P_m(U)^{I}.$$

Thus the characteristic roots of $P_m(A)$ are the same as those of $P_m(B)$, namely the products $a_1^{i_1}\ldots a_\alpha^{i_\alpha}$.

45. Adjugates. Let A be a matrix of order n, and let its r-rowed minor determinants be represented by $a_{j_1\ldots j_r}^{i_1\ldots i_r}$. The matrix of order $m = \binom{n}{r}$ whose elements are the numbers $a_{j_1\ldots j_r}^{i_1\ldots i_r}$ in the lexacographic

[1] HUNYADY, E.: J. reine angew. Math. Vol. 89 (1880) pp. 47–69.

[2] IGEL, B.: Mh. Math. Phys. Vol. 3 (1892) pp. 55–67.

[3] ESCHERICH, G. VON: Mh. Math. Phys. Vol. 3 (1892) pp. 68–80.

[4] FRANKLIN, F.: Amer. J. Math. Vol. 16 (1894) pp. 205–206. — SCHUR, I.: Über eine Klasse von Matrizen, die sich einer gegebenen Matrix zuordnen lassen, p. 17. Dissertation. Berlin 1901.

order is called the *r-th adjugate* $C_r(A)$ of A. The first adjugate of A is A itself. The $(n-1)$-th adjugate differs from A^{A} only by a permutation of the rows and columns. The n-th adjugate of A is $d(A)$.

Theorem 45.1.

$$dC_r(A) = d(A)^k, \qquad k = \binom{n-1}{r-1}.$$ [1]

Theorem 45.2.

$$C_r(AB) = C_r(A)\,C_r(B).$$

The proofs of these theorems can be made similarly to those of Theorems 44.3 and 44.2 by using as the transformable system the r-rowed minor determinants of the matrix

$$\left\| \begin{matrix} x_1^{(1)} & x_1^{(2)} & \dots & x_1^{(n)} \\ \cdot & \cdot & \cdot & \cdot \\ x_n^{(1)} & x_n^{(2)} & \dots & x_n^{(n)} \end{matrix} \right\|$$

whose columns are n independent sets of variables which are transformed cogrediently by the transformation[2]

$$x_i^{(k)\prime} = \sum a_{ij}\, x_j^{(k)} \qquad A = (a_{rs}).$$

In fact, Theorem 45.2 is but a restatement of Theorem 7.9.

Theorem 45.3. The characteristic roots of $C_r(A)$ are the $\binom{n}{r}$ products of the characteristic roots of A taken r at a time[3].

The proof can be made similarly to that of Theorem 44.4.

O. Niccoletti[4] proved that if A is of rank $\varrho \geqq r$, then $C_r(A)$ is of rank $\binom{\varrho}{r}$.

W. Burnside[5] gave another proof of Theorem 45.3.

J. Williamson[6] proved that if A has the characteristic roots $a_1, \dots, a_\alpha$, and if C is a matrix composed of m^2 blocks $f_{rs}(A)$, then the characteristic roots of C are the $m\alpha$ characteristic roots of the α m-rowed matrices $(f_{rs}(a_k))$.

The following comments were made by K. Hensel[7]: "There are associated with the ring $\mathfrak{R}[A, B, \dots]$ of all n-th order matrices, n derived rings

$$\mathfrak{R}[C_1(A), C_1(B), \dots], \qquad \mathfrak{R}[C_2(A), C_2(B), \dots], \quad \dots$$

which are isomorphic with $\mathfrak{R}$ under multiplication, and of which the last is identical with the determinants of the matrices $A, B, \dots$ All

---

1 Franke, E.: J. reine angew. Math. Vol. 61 (1863) pp. 350–356.

2 Schur, I.: l. c. Chap. II.

3 Metzler, W. H.: Amer. J. Math. Vol. 16 (1893) pp. 131–150. — Rados, G.: Math. Ann. Vol. 48 (1897) pp. 417–424.

4 Niccoletti, O.: Atti Accad. Sci. Torino Vol. 37 (1901–1902) pp. 655–659.

5 Burnside, W.: Quart. J. Math. Vol. 33 (1902) pp. 80–84.

6 Williamson, J.: Bull. Amer. Math. Soc. Vol. 37 (1931) pp. 585–590.

7 Hensel, K.: J. reine angew. Math. Vol. 159 (1928) pp. 246–254.

properties of the matrices of $\mathfrak{R}[A, B, \ldots]$ in so far as they relate only to their combination and decomposition by multiplication ... have unique counterparts in each of the derived rings. Thus if A is a divisor of zero, so are $C_2(A)$, $C_3(A)$, ... If A is a divisor of one, so are $C_2(A)$, $C_3(A)$, ... To every equivalence relation under multiplication, and to every division into classes in $\mathfrak{R}$ based on such a relation, correspond the same equivalence relation and the same class division in each derived ring. ... In order to find the complete system of invariant classes for a class division in $\mathfrak{R}(A, B, \ldots)$, it is sufficient in very many cases to seek out the most trivial invariants for the same class division for all derived rings. For these taken together furnish the complete system of invariants for $\mathfrak{R}$."

J. WILLIAMSON[1] defined REISS's matrix to be a matrix of order $\binom{n}{r}$ whose elements in any row are the determinants of the matrices obtained by replacing all sets of r columns of X^{T} by a definite set of r columns of Y^{T}, and whose elements in any column are the determinants of the matrices obtained by replacing one set of r columns of X^{T} by the $\binom{n}{r}$ sets of r columns of Y^{T} in turn. He investigated the characteristic roots of this matrix.

I. SCHUR[2] made an exhaustive investigation of invariant matrices. If A and B have independent elements, and if the elements of $T(A)$ are rational integral functions of the elements of A, and if $T(A)\,T(B) = T(AB)$, then $T(A)$ is an invariant matric function of A. Product-matrices (Theorem 44.1), power-matrices (Theorem 44.2), and adjugates (Theorem 45.2) are instances of invariant matric functions. SCHUR called a matrix *irreducible* if it is not similar to a direct sum of matrices of lower order, and proved that every invariant matric function $T(A)$ can be uniquely expressed similar to a direct sum of irreducible invariant matric functions each of which is homogeneous in the elements of A. Two invariant matrices $T_1(A)$ and $T_2(A)$ are similar if and only if they have equal traces. If $T(A)$ is of degree m in the elements of A, the characteristic roots of $T(A)$ are the products m at a time of the characteristic roots of A. If A has linear elementary divisors, the same is true of $T(A)$.

The theory of invariant matrices has had considerable development in connection with the theory of group characteristics, where it is of prime importance[3].

---

[1] WILLIAMSON, J.: Proc. Edinburgh Math. Soc. II Vol. 2 (1929) pp. 240 to 251.

[2] SCHUR, I.: Dissertation. Berlin 1901.

[3] WEYL, H.: Gruppentheorie und Quantenmechanik, p. 100. Leipzig 1928. See also the forthcoming number of this Ergebnisse entitled „Gruppentheorie" by B. L. VAN DER WAERDEN.

VIII. Matric equations.

46. The general linear equation. If $A_1, A_2, \ldots, A_h, B_1, B_2, \ldots, B_h, C$ are matrices of order n with elements in a field $\mathfrak{F}$, the general linear equation is of the form

$$(46.1) \qquad A_1 X B_1 + A_2 X B_2 + \cdots + A_h X B_h = C,$$

where X is a matrix of order n, with elements in $\mathfrak{F}$, to be found. By replacing C by 0 we obtain the corresponding *auxiliary equation*. It is evident that if X_1 and X_2 are solutions of (46.1), their difference is a solution of the auxiliary equation. Hence the sum of a particular solution of (46.1) and the general solution of the corresponding auxiliary equation gives the general solution of (46.1).

Equation (46.1) may be considered as a system of n^2 equations for the n^2 elements x_{rs} of X. The matrix of this system of equations, if the x_{rs} are arranged in the proper order, is

$$(46.2) \qquad \| A_1 \cdot \times B_1 + A_2 \cdot \times B_2 + A_n \cdot \times B_n \|.$$

The theory of systems of linear equations now gives

Theorem 46.1. A necessary and sufficient condition in order that (46.1) have a solution X_0 is that the rank ϱ of (46.2) be the same as the rank of the $n^2 \times (n^2 + 1)$ array obtained by bordering (46.2) with the elements of C. The general solution of (46.1) is then of the form

$$X = X_0 + \mu_1 X_1 + \cdots + \mu_{n^2-\varrho} X_{n^2-\varrho},$$

where $(X_1, \ldots, X_{n^2-\varrho})$ is a fundamental system of solutions of the auxiliary equation, and $\mu_1, \ldots$ vary independently over $\mathfrak{F}$.

This equation was first studied by SYLVESTER[1]. The matrix (46.2) was called by him the *nivellateur*, although he did not recognize it as a sum of direct products.

J. H. M. WEDDERBURN[2] gave a method of solution by infinite series. F. L. HITCHCOCK[3] applied GIBBS' dyadics to obtain the coefficients of the characteristic equation of (46.2), and thus to solve (46.1).

In the special case

$$(46.3) \qquad AX = B$$

there is the unique solution $A^1 B$ if $d(A) \neq 0$. Since (46.3) is equivalent to the n systems of n linear equations each

$$\sum a_{ri} x_{is} = b_{rs}, \qquad (r, s = 1, 2, \ldots, n)$$

[1] SYLVESTER: C. R. Acad. Sci., Paris Vol. 99 (1884) pp. 117–118, 409–412, 432–436 and 527–529.

[2] WEDDERBURN, J. H. M.: Proc. Edinburgh Math. Soc. Vol. 22 (1904) pp. 49 to 53.

[3] HITCHCOCK, F. L.: Proc. Nat. Acad. Sci. U.S.A. Vol. 8 (1922) pp. 78–83.

it has a solution if and only if the $n \times 2n$ array (A, B) has the same rank as A.[1]

A necessary condition that $AX = B$ and $XC = D$ have a common solution is that $AD = BC$, for $AXC = BC = AD$. If either A or B is non-singular, the condition is also sufficient[2].

The equation

$$AX = XB \tag{46.4}$$

was first discussed (merely for quaternions) by CAYLEY[3]. A necessary and sufficient condition that it have a non-singular solution is that $\lambda I - A$ and $\lambda I - B$ have the same invariant factors (Theorem 39.1). On the other hand, $X = 0$ is always a solution. By Theorem 46.1 the nivellateur is

$$\| A \cdot\times I - I \cdot\times B \|,$$

whose determinant is, by Corollary 43.82, the resultant of the characteristic equations of A and B. Thus:

Theorem 46.2. The equation $AX = XB$ has a non-zero solution if and only if A and B have a characteristic root in common[4].

The following important theorem, which generalizes the preceding theorem and gives a complete solution of (46.4), seems to have been discovered independently by CECIONI[5] and FROBENIUS[6]. The following proof is adapted from that of FROBENIUS.

Theorem 46.3. The number of linearly independent solutions of the equation $AX = XB$ is $\sum e_{ij}$ where e_{ij} is the degree of the greatest common divisor of the invariant factor a_i of $\lambda I - A$ and the invariant factor b_j of $\lambda I - B$.

Let

$$\lambda I - B = B_1, \qquad \lambda I - A = B_2.$$

Let L_1, L_2, M_1, M_2 be unimodular matrices, at present undetermined. Let

$$L_1 B_1 M_1^{\mathrm{I}} = A_1, \qquad L_2 B_2 M_2^{\mathrm{I}} = A_2.$$

Let P and Q be any two matrices whose elements are polynomials in λ such that

$$PA_1 = A_2 Q. \tag{46.5}$$

Then

$$P(L_1 B_1 M_1^{\mathrm{I}}) = (L_2 B_2 M_2^{\mathrm{I}}) Q,$$
$$(L_2^{\mathrm{I}} P L_1) B_1 = B_2 (M_2^{\mathrm{I}} Q M_1).$$

---

[1] FROBENIUS: J. reine angew. Math. Vol. 84 (1878) p. 8. — AUTONNE, L.: Ann. Univ. Lyon II Vol. 25 (1909) pp. 1—79.

[2] CECIONI, F.: Ann. Scuola norm. super. Pisa Vol. 11 (1909) pp. 1—140.

[3] CAYLEY: Mess. Math. Vol. 14 (1885) pp. 108—112.

[4] SYLVESTER: C. R. Acad. Sci., Paris Vol. 99 (1884) pp. 67—71 and 115—116. — CECIONI: Ann. Scuola norm. super. Pisa Vol. 11 (1909) pp. 1—40.

[5] CECIONI: Atti Accad. naz. Lincei, Rend. V Vol. 18[1] (1909) pp. 566—571.

[6] FROBENIUS: S.-B. preuß. Akad. Wiss. 1910 pp. 3—15.

Since the coefficients of the highest powers of λ in B_1 and B_2 are non-singular, we can by Theorem 29.1 determine unique matrices U_1, U_2, R_1, R_2 such that

$$L_2^{\mathrm{I}} P L_1 = B_2 U_2 + R_2,$$
$$M_2^{\mathrm{I}} Q M_1 = U_1 B_1 + R_1,$$

where R_2 and R_1 have elements free from λ. Then

$$(B_2 U_2 + R_2) B_1 = B_2 (U_1 B_1 + R_1),$$
$$B_2 (U_2 - U_1) B_1 = B_2 R_1 - R_2 B_1.$$

If $U_2 - U_1$ were not O, the left side would be of at least the second degree in λ and the right side of at most the first degree. Hence $U_2 = U_1$ and

$$(\lambda I - A) R_1 = R_2 (\lambda I - B).$$

Since A, B, R_1, R_2 are independent of λ,

$$R_1 = R_2 \equiv R, \qquad AR = RB.$$

From every pair of matrices P and Q which satisfy (46.5) there arise, if L and M are fixed, a unique constant matrix R satisfying (46.4).

If, conversely, R is any matrix such that $AR = RB$, then $RB_1 = B_2 R$. Let U be an arbitrary matrix whose elements are polynomials in λ, and set

$$P = L_2 (B_2 U + R) L_1^{\mathrm{I}},$$
$$Q = M_2 (U B_1 + R) M_1^{\mathrm{I}},$$

or

$$L_2^{\mathrm{I}} P L_1 = B_2 U + R, \qquad M_2^{\mathrm{I}} Q M_1 = U B_1 + R.$$

Then

$$(L_2^{\mathrm{I}} P L_1) B_1 = B_2 (M_2^{\mathrm{I}} Q M_1),$$
$$P (L_1 B_1 M_1^{\mathrm{I}}) = (L_2 B_2 M_2^{\mathrm{I}}) Q,$$

and hence $PA_1 = A_2 Q$.

To the matrix R correspond infinitely many pairs P, Q. If P, Q is a definite pair obtained by means of the matrix U, and if $P - P_0$, $Q - Q_0$ are any others obtained by means of $U - U_0$, then

$$L_2^{\mathrm{I}} P_0 L_1 = B_2 U_0 = (L_2^{\mathrm{I}} A_2 M_2) U_0,$$
$$M_2^{\mathrm{I}} Q_0 M_1 = U_0 B_1 = U_0 (L_1^{\mathrm{I}} A_1 M_1),$$

or if we set $M_2 U_0 L_1^{\mathrm{I}} = T$,

$$P_0 = A_2 T, \qquad Q_0 = T A_1.$$

If therefore T is an arbitrary matrix,

$$P - A_2 T, \qquad Q - T A_1$$

is the most general pair of matrices which corresponds to the constant matrix R such that $RB_1 = B_2 R$.

Now let $A_1 = (a_{1rs})$, $A_2 = (a_{2rs})$ be the normal forms of B_1 and B_2 respectively. Then $a_{1rs} = \delta_{rs} b_r$, $a_{2rs} = \delta_{rs} a_r$ where a_r and b_r are the invariant factors of A and B respectively (§ 27). The condition $PA_1 = A_2 Q$ gives

$$p_{rs} b_s = a_r q_{rs}, \qquad P = (p_{rs}), \quad Q = (q_{rs})$$

or

$$p_{rs} = \frac{a_r}{b_s} q_{rs}.$$

The elements of $P - A_2 T$ and $Q - TA_1$ are therefore

$$p_{rs} - a_r t_{rs} = \frac{a_r}{b_s}(q_{rs} - b_s t_{rs}), \qquad T = (t_{rs}).$$

By a suitable choice of t_{rs}, the degree of $q_{rs} - b_s t_{rs}$ can be made less than the degree of b_s. Then t_{rs} and $q_{rs} - b_r t_{rs}$ are uniquely determined. Let e_{rs} be the degree of the greatest common divisor d_{rs} of a_r and b_s. In order to obtain the most general $p_{rs} - a_r t_{rs}$ which is a polynomial, it is necessary and sufficient that $q_{rs} - b_s t_{rs}$ be the product of b_s/d_{rs} by an arbitrary polynomial of degree $e_{rs} - 1$ — that is, a polynomial with e_{rs} arbitrary coefficients. Hence there are exactly $\sum e_{rs}$ linearly independent matrices P (and for each P a unique Q), each of which corresponds to a unique R. Hence the number of linearly independent matrices R satisfying (46.4) is $\sum e_{rs}$.

The solution of the equation $AX = XB$ has been discussed by G. LANDSBERG[1], R. WILSON[2], and H. W. TURNBULL and A. C. AITKEN[3].

Corollary 46.31. In a field $\mathfrak{F}$ containing the elements of A, B and C, the equation

$$AX + XB = C$$

has no solution unless the rank of $A \cdot \times B$ is the same as the rank of this matrix augmented with the elements of C. If these ranks are the same, every solution is of the form

$$X_0 + \sum \lambda_i X_i$$

where X_0 is a particular solution of the given equation, the X_i are $\sum e_{rs}$ linearly independent solutions of the auxiliary equation $AX + XB = 0$, the λ_i are arbitrary in $\mathfrak{F}$, and e_{rs} is the degree of the g.c.d. of the invariant factor a_r of $\lambda I - A$ and the invariant factor b_s of $\lambda I - B$.

This equation has been discussed by CECIONI[4], D. E. RUTHERFORD[5], and R. WEITZENBÖCK[6].

---

[1] LANDSBERG, G.: J. reine angew. Math. Vol. 116 (1896) pp. 331–349.

[2] WILSON, R.: Proc. London Math. Soc. II Vol. 30 (1930) pp. 359–366; Vol. 33 (1932) pp. 517–524.

[3] TURNBULL, H. W., and A. C. AITKEN: Canonical matrices, Chap. X. London 1932.

[4] CECIONI: Ann. Scuola norm. super. Pisa Vol. 11 (1909) pp. 1–40.

[5] RUTHERFORD, D. E.: Akad. Wetensch. Amsterdam, Proc. Vol. 35 (1932) pp. 54–59.

[6] WEITZENBÖCK, R.: Akad. Wetensch. Amsterdam, Proc. Vol. 35 (1932) pp. 60 to 61.

Corollary 46.32. The number of linearly independent solutions of the equation $AX = XA$ is

$$n + 2(m_1 + \cdots + m_n),$$

where m_i is the degree of the g.c.d. d_i of the i-rowed minor determinants of $\lambda I - A$.

Let e_r be the degree of the invariant factor a_r of $\lambda I - A$. Let e_{rs} be the degree of the g.c.d. of a_r and a_s. Then

$$e_{rr} = e_r, \quad e_{rs} = e_{sr} = e_s, \quad r > s.$$

Hence

$$\begin{aligned}\sum e_{rs} = e_1 + e_2 + \cdots + e_n + 2[e_1 + (e_1 + e_2) + (e_1 + e_2 + e_3) + \cdots \\ + (e_1 + e_2 + e_{n-1})] \\ = n + 2(m_1 + \cdots + m_n)\end{aligned}$$

by Theorem 27.1.

This theorem was stated by FROBENIUS[1] and proved by MAURER[2], A. VOSS[3], G. LANDSBERG[4], K. HENSEL[5], F. CECIONI[6], and U. AMALDI[7].

W. K. CLIFFORD[8] attempted to prove that every matrix commutative with A is a polynomial in A. SYLVESTER[9] showed that this is not so.

H. LAURENT[10] gave a false proof that if $AB = BA$, then both A and B are polynomials in the same matrix. An example to show that this is not always so was given by H. B. PHILLIPS[11].

A. VOSS[12] proved that if $AP = QA$, where $P = P_1 \dot{+} P_2$, $Q = Q_1 \dot{+} Q_2$, P_1 and Q_1 of the same order, then $A = A_1 \dot{+} A_2$ where A_1 is of the same order as P_1.

J. PLEMELJ[13] proved that if certain matrices A_i are commutative in pairs, a matrix T exists such that $T^I A_i T$ are simultaneously equal to a direct sum of matrices having equal characteristic roots.

H. TABER[14] wrote out explicitly the most general matrix commutative with A in terms of its characteristic roots, and later[15] "proved the

---

1 FROBENIUS: J. reine angew. Math. Vol. 84 (1878) pp. 1—63.

2 MAURER: Zur Theorie der linearen Substitutionen. Dissertation. Straßburg 1887.

3 VOSS, A.: S.-B. Bayer. Akad. Wiss. Vol. 19 (1889) pp. 283—300.

4 LANDSBERG, G.: J. reine angew. Math. Vol. 116 (1896) pp. 331—349.

5 HENSEL, K.: J. reine angew. Math. Vol. 127 (1904) pp. 116—166.

6 CECIONI, F.: Ann. Scuola norm. super. Pisa Vol. 11 (1909) pp. 1—140.

7 AMALDI, U.: Ist. Lombardo, Rend. II Vol. 45 (1912) pp. 433—445.

8 CLIFFORD, W. K.: Fragment on matrices. Collected Papers, p. 337. 1875.

9 SYLVESTER: John Hopkins Circ. Vol. 3 (1884) pp. 33, 34, 57.

10 LAURENT, H.: J. Math. pures appl. V Vol. 4 (1898) pp. 75—119.

11 PHILLIPS, H. B.: Amer. J. Math. Vol. 41 (1919) pp. 256—278.

12 VOSS, A.: S.-B. Bayer. Akad. Wiss. II Vol. 17 (1892) pp. 235—356.

13 PLEMELJ, J.: Mh. Math. Phys. Vol. 12 (1901) pp. 82—96.

14 TABER, H.: Proc. Amer. Acad. Arts Sci. Vol. 26 (1890—1891) pp. 64—66.

15 TABER, H.: Proc. Amer. Acad. Arts Sci. Vol. 27 (1891—1892) pp. 46—56.

surmise of SYLVESTER[1] that if A is not derogatory (§ 15) the only matrices commutative with A are polynomial functions of A". An equivalent theorem had, however, been previously given by FROBENIUS[2], namely that if the first minors of the characteristic determinant of A are relatively prime, the only matrices commutative with A are polynomials in A. (See Theorem 15.1.)

Explicit solutions of $AX = XB$ were given by H. KREIS[3] and L. AUTONNE[4]. Numerical examples of commutative matrices were given by H. HILTON[5].

M. KRAWTCHOUK[6] found the number of linearly independent matrices in a commutative group of matrices.

I. SCHUR[7] proved that the order of a commutative group of n-th order matrices is $\leqq [n^2/4 + 1]$.

A. RANUM[8] found necessary and sufficient conditions that a singular matrix belong to a group.

47. Scalar equations. Let

$$p(X) = p_0 X^m + p_1 X^{m-1} + \cdots + p_m I = O \tag{47.1}$$

be an equation of degree m with coefficients in a field $\mathfrak{F}$, where X is an n-th order matrix to be determined. If X_1 is a solution with elements in $\mathfrak{F}$, so is $X_2 = P^{\mathrm{I}} X_1 P$ where P is an arbitrary non-singular matrix with elements in $\mathfrak{F}$. Hence all solutions are determined by those in canonical form.

If X satisfies $p(\lambda) = 0$, the minimum equation of X divides $p(\lambda)$ and conversely (Theorem 13.1). Equation (47.1) is completely solved, then, by finding the factors of $p(\lambda)$ which are irreducible in $\mathfrak{F}$, and setting up those matrices $X_1, X_2, \ldots, X_k$ in canonical form (Corollary 39.12) whose minimum equations divide $p(\lambda)$. The number of dissimilar solutions is finite.

SYLVESTER[9] discussed the equations $X^m = I$ and $X^m = O$. There exists an integer m such that $A^m = I$ if and only if the characteristic roots of A are roots of unity[10], in which event the elementary divisors of A are simple (§ 26)[11].

A necessary and sufficient condition that there exist an integer $m > 1$ such that $A^m = A$, whether A be singular or not, is that the

[1] SYLVESTER: C. R. Acad. Sci., Paris Vol. 98 (1884) p. 471.
[2] FROBENIUS: J. reine angew. Math. Vol. 84 (1878) pp. 1—63 Theorem XIII.
[3] KREIS, H.: Contribution à la théorie des systèmes linéaires. Zürich 1906.
[4] AUTONNE, L.: J. École polytechn. Vol. 14 (1910) pp. 83—131.
[5] HILTON, H.: Mess. Math. Vol. 41 (1911) pp. 110—118.
[6] KRAWTCHOUK, M.: Rend. Circ. mat. Palermo Vol. 51 (1927) pp. 126—130.
[7] SCHUR, I.: J. reine angew. Math. Vol. 130 (1905) pp. 66—76.
[8] RANUM, A.: Amer. J. Math. Vol. 31 (1909) pp. 18—41.
[9] SYLVESTER: C. R. Acad. Sci., Paris Vol. 94 (1882) pp. 396—399.
[10] LIPSCHITZ: Acta math. Vol. 10 (1887) pp. 137—144.
[11] BAKER, H. F.: Proc. London Math. Soc. Vol. 35 (1903) pp. 379—384.

characteristic roots of A be 0 or roots of unity, and that the elementary divisors of A be simple[1].

H. W. TURNBULL[2] found that if

$$x_{rs} = (-1)^{n-s}\binom{n-s}{r-1},$$

then $X^3 = (x_{rs})^3 = I$.

R. VAIDYANATHASWAMY[3] proved that if $r = 2^m p_1^{e_1} \ldots p_\lambda^{e_\lambda}$ where the p's are distinct odd primes, a necessary and sufficient condition for the existence of an integral matrix of order n and period r is that

$$\varphi(2^m) + \varphi(p_1^{e_1}) + \cdots + \varphi(p_\lambda^{e_\lambda}) \leqq n, \qquad m > 1,$$
$$\varphi(p_1^{e_1}) + \cdots + \varphi(p_\lambda^{e_\lambda}) \leqq n, \qquad m = 0, 1,$$

where φ is the totient.

48. The unilateral equation. We now consider the equation

$$(48.1) \quad F(X) = A_0 X^m + A_1 X^{m-1} + \cdots + A_{m-1} X + A_m = O,$$

where the A_i are n-th order matrices with elements in $\mathfrak{F}$. Let $F(\lambda)$ denote the matrix

$$A_0\lambda^m + A_1\lambda^{m-1} + \cdots + A_{m-1}\lambda + A_m,$$

where λ is indeterminate. Then by the theorem of PHILLIPS (Theorem 14.2), X satisfies the scalar equation

$$(48.2) \qquad f(\lambda) = dF(\lambda) = 0$$

which, unless it vanish identically, is of degree $\leqq nm$.

Let $Y_1, Y_2, \ldots, Y_k$ be the solutions of (48.2) determined as in § 47. Then every solution of (48.1) is of the form $X_i = P_i Y_i P_i^{\mathrm{I}}$ where P_i must be non-singular. With this substitution (48.1) becomes

$$A_0 P_i Y_i^m + A_1 P_i Y_i^{m-1} + \cdots + A_{m-1} P_i Y_i + A_m P_i = O,$$

which is a linear equation of the form (46.1) with $C = O$ for the matrix P_i.

In case $f(\lambda) \equiv 0$, there are more than a finite number of dissimilar solutions[4].

CAYLEY first discussed the equation $X^2 = A$ for matrices of orders 2 and 3.[5] SYLVESTER showed[6] that every characteristic root of a solution X of $PX^2 + QX + R = O$, where all matrices are of the second order, is a root of $|P\lambda^2 + Q\lambda + R| = 0$. He made several attempts to determine the number of solutions of the equation (48.1) for special cases[7].

---

1 RANUM, A.: Amer. J. Math. Vol. 31 (1909) pp. 18–41.

2 TURNBULL, H. W.: J. London Math. Soc. Vol. 2 (1927) pp. 242–244.

3 VAIDYANATHASWAMY, R.: J. London Math. Soc. Vol. 3 (1928) pp. 268–272.

4 ROTH, W. E.: Trans. Amer. Math. Soc. Vol. 32 (1929) pp. 61–80.

5 CAYLEY: Philos. Trans. Roy. Soc. London Vol. 148 (1858) pp. 39–46.

6 SYLVESTER: C. R. Acad. Sci., Paris Vol. 99 (1884) pp. 555–558 and 621–631.

7 SYLVESTER: Johns Hopkins Circ. Vol. 3 (1884) p. 122 — Philos. Mag. Vol. 18 (1884) pp. 454–458 — Quart. J. Math. Vol. 20 (1885) pp. 305–312 — C. R. Acad. Sci., Paris Vol. 99 (1884) pp. 13–15.

A. BUCHHEIM[1] proved in general that if X satisfies (48.1), each of its characteristic roots satisfies (48.2).

FROBENIUS[2] found all solutions of $X^2 = A$ in the complex field, when $d(A) \neq 0$, which are expressible as polynomials in A (see Lemma 35.22). The same method was extended by H. F. BAKER[3] and L. E. DICKSON[4] to find all solutions of $X^m = A$ in the complex field which are polynomials in A.

H. KREIS[5] treated the equation $p(X) = A$ where $p(\lambda)$ is a polynomial with complex coefficients, $d(A) \neq 0$, and obtained solutions which are polynomials in A. Later[6] he found necessary and sufficient conditions for the solvability of this equation when $d(A) = 0$.

W. E. ROTH[7] proved that a necessary and sufficient condition that the equation $p(X) = A$ have a solution in the complex field expressible as a polynomial in A is that the equation $p(x) = a_i$ have at least one simple root for each characteristic root of A corresponding to a non-linear elementary divisor. The number of distinct solutions of $p(X) = A$ which are expressible as polynomials in A is $\sum_{j=1}^{s} \mu_j$ where s is the number of distinct roots of the minimum equation $m(\lambda) = 0$ of A, and μ_j is the number of distinct roots of $p(\lambda) - a_j = 0$ when a_j is a simple root of $m(\lambda) = 0$, and is the number of simple roots of $p(\lambda) - a_j = 0$ when a_j is a multiple root of $m(\lambda) = 0$.

A method for finding the solutions of $p(X) = A$ which are not polynomials in A was given by P. FRANKLIN[8].

D. E. RUTHERFORD[9] gave a more explicit form of solution. Let $U_n = (\delta_{r,s-1})$ be a matrix of order n. Let $a_1, a_2, \ldots$ be the characteristic roots of A. Then $A \stackrel{S}{=} N_1 \dotplus N_2 \dotplus \cdots \dotplus N_\varrho$ where

$$N_h = C_{\xi_h}(a_h)_{\vartheta_h} \equiv a_h I_{\xi_h} + U_{\xi_h}^{\theta_h}.$$

If for any arrangement of the N's there is for every h a θ_h-fold repeated root b_h of $p(\lambda) = a_h$, then a solution

$$Y = C_{\xi_1}(b_1) \dotplus \cdots \dotplus C_{\xi_\varrho}(b_\varrho)$$

of $p(X) = A$ exists. Comparing this with ROTH's result indicates that the solution is a polynomial in A if and only if $\theta_h = 1$.

1 BUCHHEIM, A.: Proc. London Math. Soc. Vol. 16 (1884) pp. 63–82.

2 FROBENIUS: S.-B. preuß. Akad. Wiss. 1896 pp. 7–16.

3 BAKER, H. F.: Proc. Cambridge Philos. Soc. Vol. 23 (1925) pp. 22–27.

4 DICKSON, L. E.: Modern algebraic theories, p. 120. Chicago 1926.

5 KREIS, H.: Contribution à la théorie des systèmes linéaires. Zürich 1906.

6 KREIS, H.: Vjschr. naturforsch. Ges. Zürich Vol. 53 (1908) pp. 366–376.

7 ROTH, W. E.: Trans. Amer. Math. Soc. Vol. 30 (1928) pp. 579–596.

8 FRANKLIN, P.: J. Math. Physics, Massachusetts Inst. Technol. Vol. 10 (1932) pp. 289–314.

9 RUTHERFORD, D. E.: Proc. Edinburgh Math. Soc. II Vol. 3 (1932) pp. 135 to 143.

R. WEITZENBÖCK[1] showed that the method of FROBENIUS may be extended so as to yield all solutions of $X^m = A$.

W. E. ROTH[2] considered the general unilateral equation (48.1) where the A_i are $m \times n$ arrays and X an $n \times n$ matrix with elements in the complex field. Necessary conditions for a solution are given, and it is indicated how a solution, if it exists, can be computed. Several examples are given, as well as a quite complete bibliography.

A. HURWITZ[3] found the most general system of p matrices, each of order n, which satisfy the conditions

$$B_h^2 = I, \quad B_h B_k = -B_k B_h, \quad h \neq k,$$

and the number of dissimilar systems.

A. S. EDDINGTON[4] proved that if $B_1, B_2, \ldots, B_p$ are four-rowed matrices such that

$$B_h^2 = -I, \quad B_h B_k = -B_k B_h, \quad k \neq h,$$

then $p \leqq 5$. If the elements of each matrix are all real or all imaginary, then there are 2 real and 3 imaginary matrices in every set of 5.

M. H. A. NEWMAN[5] generalized EDDINGTON's result and simplified the proof. If $B_1, B_2, \ldots, B_p$ are a set of n-rowed matrices, $n = 2^q r$, r odd, and if

$$B_h^2 = -I, \quad B_h B_k = -B_k B_h, \quad k \neq h,$$

then $p \leqq 2q + 1$; and the maximum is attained. If in a maximal set there are ϱ with real elements and the remaining σ with purely imaginary elements, then $\varrho - \sigma = -1$ or 7. He extended these considerations to hermitian matrices.

IX. Functions of Matrices.

49. Power series in matrices. Let $P(\lambda) = \sum_{i=0}^{\infty} a_i \lambda^i$ be an ordinary power series with complex coefficients in the complex variable λ. If for a matrix A of order n with complex elements every element of

$$P_m(A) = \sum_{i=0}^{m} a_i A^i$$

approaches a finite limit as $m \to \infty$, the matrix

$$P(A) = \sum_{i=0}^{\infty} a_i A^i$$

[1] WEITZENBÖCK, R.: Akad. Wetensch. Amsterdam, Proc. Vol. 35 (1932) pp. 157 to 161.

[2] ROTH, W. E.: Trans. Amer. Math. Soc. Vol. 32 (1930) pp. 61–80.

[3] HURWITZ, A.: Math. Ann. Vol. 88 (1923) pp. 1–25.

[4] EDDINGTON, A. S.: J. London Math. Soc. Vol. 7 (1932) pp. 58–68.

[5] NEWMAN, M. H. A.: J. London Math. Soc. Vol. 7 (1932) pp. 93–99.

is said to exist and to be equal to the matrix of these limiting values.

Theorem 49. The power series $P(A)$ converges if and only if every characteristic root of A lies inside or on the circle of convergence of $P(\lambda)$, and for every ν-fold characteristic root λ_i which lies on the circle of convergence, the $(\nu-1)$-th derivative $P^{(\nu-1)}(\lambda_i)$ converges.

This theorem and proof are due to K. HENSEL[1]. E. WEYR[2] had previously proved the theorem for the case where no characteristic root lies on the circle of convergence.

Let us write

$$A \overset{S}{=} A_1 \dot{+} A_2 \dot{+} \cdots \dot{+} A_k,$$

where

$$A_i = \begin{Vmatrix} \lambda_i & 1 & 0 & \cdots & 0 \\ 0 & \lambda_i & 1 & \cdots & 0 \\ \cdot & \cdot & \cdot & \cdot & \cdot \\ 0 & 0 & 0 & \cdots & \lambda_i \end{Vmatrix}$$

is of order ν_i (Corollary 39.11). Then

$$P(A) \overset{S}{=} P(A_1) \dot{+} P(A_2) \dot{+} \cdots \dot{+} P(A_k),$$

so that $P(A)$ converges if and only if every $P(A_i)$ converges.

Write ν for ν_i, and let

$$A_i - \lambda_i I = U, \qquad U^{\nu-1} \neq 0, \qquad U^{\nu} = 0.$$

Then for $m \geqq \nu$

$$\begin{aligned} P_m(A_i) &= \sum_{j=0}^{m} a_j (\lambda_i I + U)^j \\ &= \sum_{j=0}^{m} a_j \sum_{h=0}^{j} \binom{j}{h} \lambda_i^{j-h} U^h \\ &= \sum_{h=0}^{m} U^h \left[\sum_{j=h}^{m} a_j \binom{j}{h} \lambda_i^{j-h}\right] \\ &= \sum_{h=0}^{\nu-1} \frac{1}{h!} P_m^{(h)}(\lambda_i)\, U^h. \end{aligned}$$

Hence

$$P(A_i) = \sum_{h=0}^{\nu-1} \frac{1}{h!} P^{(h)}(\lambda_i)\, U_h,$$

where $P^{(h)}$ denotes the h-th derivative.

---

[1] HENSEL, K.: J. reine angew. Math. Vol. 155 (1926) pp. 107–110.

[2] WEYR, E.: Bull. Sci. math. II Vol. 11 (1887) pp. 205–215.

Evidently $P(A_i)$ converges if and only if

$$P(\lambda_i),\ P'(\lambda_i),\ \ldots,\ P^{(\nu-1)}(\lambda_i)$$

all converge. If λ_i is outside the circle of convergence of $P(\lambda)$, $P(\lambda_i)$ diverges, while if λ_i is inside this circle, all derivatives converge. If λ_i is on the circle, all these derivatives converge if and only if $P^{\nu-1}(\lambda_i)$ converges.

G. PEANO[1] and E. CARVALLO[2] discussed the function e^A, and the former discussed the TAYLOR series in matrices.

H. TABER[3] defined trigonometric functions of a matrix. He also proved[4] that every real proper orthogonal matrix can be represented in the form e^A where A is skew.

W. H. METZLER[5] considered transcendental functions of a matrix.

The following results are due to H. B. PHILLIPS[6]. If $A, B, \ldots, P$ are commutative matrices whose characteristic roots $a_i, b_i, \ldots p_i$ are ordered as in Theorem 16.1, and if $f(a, b, \ldots, p)$ is an infinite series, then $f(A, B, \ldots, P)$ converges if the series $f(a_i, b_i, \ldots, p_i)$ and their partial derivatives of proper orders converge. The TAYLOR series

$$f(Z) = f(A) + f'(A)(Z - A) + \cdots + f^{(m)}(A)\frac{(Z-A)^m}{m!} + \cdots$$

is valid for a matrix Z commutative with A if each characteristic root of Z lies within a circle with center at the corresponding root of A in which $f(z)$ is analytic, z being an ordinary complex variable.

50. Functions of matrices. That the definition of analytic function of a matrix as a power series in that matrix with scalar coefficients is too restrictive can be seen at once from the example

$$A = \begin{Vmatrix} 2 & -1 \\ 3 & -2 \end{Vmatrix}, \qquad A^2 = I.$$

Surely A is an analytic function of I, yet every power series in I is scalar.

The first attempt to define an analytic function of a matrix A having distinct characteristic roots was made by SYLVESTER[7] by means of the "interpolation formula"

$$f(A) = \sum \frac{(A - \lambda_2 I)(A - \lambda_3 I) \ldots (A - \lambda_n I)}{(\lambda_1 - \lambda_2)(\lambda_1 - \lambda_3) \ldots (\lambda_1 - \lambda_n)} f(\lambda_1)$$

[1] PEANO, G.: Math. Ann. Vol. 32 (1888) pp. 450–456.

[2] CARVALLO, E.: Mh. Math. Phys. Vol. 2 (1891) pp. 177–216, 225–266 and 311–330.

[3] TABER, H.: Amer. J. Math. Vol. 12 (1890) pp. 337–396; Vol. 13 (1891) pp. 159–172.

[4] TABER, H.: Proc. Amer. Acad. Arts Sci. Vol. 27 (1891–1892) pp. 163–165.

[5] METZLER, W. H.: Amer. J. Math. Vol. 14 (1892) pp. 326–377.

[6] PHILLIPS, H. B.: Amer. J. Math. Vol. 41 (1919) pp. 266–278.

[7] SYLVESTER: Philos. Mag. Vol. 16 (1883) pp. 267–269.

summed over all characteristic roots. The case where the roots are not distinct was considered by A. BUCHHEIM[1]. Evidently this definition yields only polynomials in A.

P. A. M. DIRAC[2] proposed to define as an analytic function of A any matrix which is commutative with all matrices which are commutative with A. This again yields only polynomials in A.[3]

L. FANTAPPIÉ[4] suggested that a satisfactory definition of analytic function of a matrix must satisfy the following conditions. If $f(x)$ and $g(x)$ are functions of a function field (real or complex) and if $f(A)$ and $g(A)$ are analytic functions of a matrix A, there must exist a correspondance $\backsim$ such that, if $f(A) \backsim f(x)$ and $g(A) \backsim g(x)$, then

1. $f(A) + g(A) \backsim f(x) + g(x)$,
2. $f(A) g(A) \backsim f(x) g(x)$,
3. $f(x) = k \backsim kI$ and $f(x) = x \backsim A$,
4. If $f(x)$ is analytic in a parameter t, and if $f(x, t) \backsim f(A, t)$, then the elements of $f(A, t)$ depend analytically upon t.

He then proved that the elements f_{rs} of $f(A)$ are the sums of the residues of $-\frac{D_{sr}(t)}{D(t)} f(t)$ at the characteristic roots of A. Here $D_{sr}(t)$ is the cofactor of f_{sr}, and $D(t) = d(A)$. If λ_i is a characteristic root of A of multiplicity ν_i, the elements f_{rs} of $f(A)$ may be expressed as linear combinations of the values at the points $x = \lambda_i$ of the function $f(x)$ and its derivatives up to order $\nu_i - 1$ at most. The coefficients depend only upon A.

G. GIORGI[5] suggested the following definition. If

$$A = P^{\mathrm{I}}(A_1 \dotplus \cdots \dotplus A_k) P,$$

where

$$A_i = \begin{Vmatrix} \lambda_i & 1 & 0 & \cdots \\ 0 & \lambda_i & 1 & \cdots \\ 0 & 0 & \lambda_i & \cdots \\ \cdot & \cdot & \cdot & \cdot \end{Vmatrix}, \qquad f(A_i) = \begin{Vmatrix} f(\lambda_i) & f'(\lambda_i) & \frac{1}{2!} f''(\lambda_i) & \cdots \\ 0 & f(\lambda_i) & f'(\lambda_i) & \cdots \\ 0 & 0 & f(\lambda_i) & \cdots \\ \cdot & \cdot & \cdot & \cdot \end{Vmatrix},$$

then

$$f(A) = P^{\mathrm{I}}[f(A_1) \dotplus \cdots \dotplus f(A_k)] P.$$

He also stated that the definition

$$f(A) = \frac{1}{2\pi i} \int \frac{f(x) I}{xI - A} dx$$

had been suggested to him in a letter by CARTAN. The formula means that the element in the (r, s)-position in $f(A)$ is the integral around a

[1] BUCHHEIM, A.: Philos. Mag. V Vol. 22 (1886) pp. 173—174.

[2] DIRAC, P. A. M.: Proc. Cambridge Philos. Soc. Vol. 23 (1926) pp. 412—418.

[3] TURNBULL and AITKEN: Canonical matrices, p. 150. London 1932.

[4] FANTAPPIÉ, L.: C. R. Acad. Sci., Paris Vol. 186 (1928) pp. 619—621.

[5] GIORGI, G.: Atti Accad. naz. Lincei, Rend. VI Vol. 8 (1928) pp. 3—8.

small closed path of the element in the (r, s)-position in the matrix $f(x)(xI - A)^{I}$.

Neither of these definitions takes care of the example given at the beginning of this section.

What seems to be the most satisfactory definition so far proposed for a multiple-valued analytic function of a matrix is the one given by M. CIPOLLA[1]. It is an extension of the definition of GIORGI, and differs from it only in the respect that different determinations for f may be used in the matrices $f(A_1), \ldots, f(A_k)$, and P must range over all matrices such that $A = P^{I}(A_1 \dotplus \cdots)P$ holds. If the same determination for f is used throughout, a *principal value* of $f(A)$ results. This definition and this only is sufficiently broad to include as functions of I all solutions of the matric equation $X^2 = I$.

Explicit results in certain special cases have been obtained. Thus M. BOTASSO[2] found A^n explicitly when the minimum equation of A is quadratic.

S. MARTIS-BIDDAU[3] gave an explicit form of A^n where A is of the second order according to FANTAPPIÉ's definition of function. Later[4] she treated e^{tA} for A of the second order by GIORGI's definition, and[5] the function A^n where A is of the third order.

E. PORCU-TORTRINI[6] gave $f(A)$ explicitly for A of the second order according to GIORGI's definition of function.

S. AMANTE[7] used FANTAPPIÉ's results to solve $f(X) = O$ where $f(z)$ is a complex function.

51. Matrices whose elements are functions of complex variables. To consider this topic in detail would take us into the theory of differential equations. Only a few results of interest in pure matric theory will be given.

J. H. M. WEDDERBURN[8] considered matric functions—i.e., matrices whose elements are analytic functions of a single complex variable. His principal result (the analog of Theorem 26.2) is that if $A(\lambda)$ is such a matric function of rank r which is holomorphic in a region $\mathfrak{R}$, there

---

1 CIPOLLA, M.: Rend. Circ. mat. Palermo Vol. 56 (1932) pp. 144–154.

2 BOTASSO, M.: Rend. Circ. mat. Palermo Vol. 35 (1913) pp. 1–46.

3 MARTIS-BIDDAU, S.: Atti Accad. naz. Lincei, Rend. VI Vol. 8 (1928) pp. 130 to 133.

4 MARTIS-BIDDAU, S.: Atti Accad. naz. Lincei, Rend. VI Vol. 8 (1928) pp. 276 to 280.

5 MARTIS-BIDDAU, S.: Atti Accad. naz. Lincei, Rend. VI Vol. 9 (1929) pp. 206 to 213.

6 PORCU-TORTRINI, E.: Atti Accad. naz. Lincei, Rend. VI Vol. 7 (1928) pp. 206 to 208.

7 AMANTE, S.: Atti Accad. naz. Lincei, Rend. VI Vol. 12 (1930) p. 290.

8 WEDDERBURN, J. H. M.: Trans. Amer. Math. Soc. Vol. 16 (1915) pp. 328 to 332.

exist two matrix functions $P(\lambda)$ and $Q(\lambda)$, which are holomorphic and non-singular in $\mathfrak{R}$, and are such that

$$P(\lambda)A(\lambda)Q(\lambda) = E_1(\lambda) \dotplus E_2(\lambda) \dotplus \cdots \dotplus E_r(\lambda) \dotplus O,$$

where $E_1(\lambda), \ldots, E_r(\lambda)$ are functions of λ which are holomorphic in $\mathfrak{R}$ and are such that E_s is a factor of E_t when $s < t$.

G. D. BIRKHOFF[1] gave the following theorem. Let $L = \|l_{rs}(x)\|$ be a matrix of functions which are single-valued and analytic for $|x| \geqq R$ (but not necessarily at $x = \infty$), and such that $d(L) \neq 0$ for $|x| \geqq R$. Then there is a matrix $A = \|a_{rs}(x)\|$ of functions analytic at $x = \infty$ reducing to I at $x = \infty$, and a matrix $E = \|e_{rs}(x)\|$ of entire functions of determinant nowhere 0 in the finite plane such that $L = A \cdot \|e_{rs}(x)\, x^{k_s}\|$ where the k's are integers.

BIRKHOFF[2] considered infinite products of analytic matrices. If every element of $A(x)$ and $B(x)$ are analytic near $x = x_0$ but perhaps not at $x = x_0$, and if $M(x)$ has elements analytic at $x = x_0$, and if $A(x) = M(x)\, B(x)$, then $A(x)$ is *left-equivalent* to $B(x)$ at $x = x_0$. The relation of left-equivalence is determinative, reflexive, symmetric and transitive. A matrix $U_i(x)$ is *elementary* if it is the identity matrix with the i-th column replaced by $c_1, c_2, \ldots, c_{i-1}, x - x_0, c_{i+1}, \ldots, c_n$. The following equivalence problem is considered. Given polynomial matrices, each having its only finite singular point at x_i (these being assumed distinct), to construct a polynomial matrix $P(x)$ equivalent to these matrices at x_i and having no other finite singular point. The most general solution is

$$P_0(x)\, U_1(x)\, U_2(x) \ldots U_n(x),$$

where $U_i(x)$ is an elementary matrix with a singular point at $x = x_i$, and $P_0(x)$ is any polynomial matrix of order 0.

The *absolute value* of a matrix whose elements are complex numbers or functions was defined by J. H. M. WEDDERBURN[3] to be

$$\lfloor A \rceil = \left(\sum a_{pq}\, \bar{a}_{pq}\right)^{\frac{1}{2}} \qquad A = (a_{rs}).$$

The following inequalities were obtained, λ being scalar.

$$\lfloor A + B \rceil \leqq \lfloor A \rceil + \lfloor B \rceil, \qquad \lfloor \lambda \rceil = n^{\frac{1}{2}} |\lambda|,$$
$$\lfloor \lambda A \rceil = |\lambda| \lfloor A \rceil, \qquad \lfloor AB \rceil \leqq \lfloor A \rceil \lfloor B \rceil.$$

52. Derivatives and integrals of matrices. These concepts were first considered by V. VOLTERRA[4]. Let $S(x)$ be a matrix of order n whose

[1] BIRKHOFF, G. D.: Math. Ann Vol. 74 (1913) pp. 122–133.

[2] BIRKHOFF: Trans. Amer. Math. Soc. Vol. 17 (1916) pp. 386–404.

[3] WEDDERBURN, J. H. M.: Bull. Amer. Math. Soc. Vol. 31 (1925) pp. 304 to 308.

[4] VOLTERRA, V.: Atti Accad. naz. Lincei, Rend. IV Vol. 3[1] (1887) pp. 393–396.

elements are finite continuous functions of the real or complex variable x. Assume $dS(x) \neq 0$. If the two limits

$$\lim_{\Delta x \to 0} \frac{S^{I}(x)\, S(x+\Delta x) - I}{\Delta x}, \qquad \lim_{\Delta x \to 0} \frac{S(x+\Delta x)\, S^{I}(x) - I}{\Delta x}$$

exist, they are called the *right* and *left derivatives* of $S(x)$, respectively.

It is to be noted, however, that since $I = S^{I}(x)\, S(x) = S(x)\, S^{I}(x)$, the right and left derivatives are merely $S^{I}(x)\, S^{D}(x)$ and $S^{D}(x)\, S^{I}(x)$, respectively, where $S^{D}(x)$ is obtained from $S(x)$ by replacing each element by its derivative.

Let $T(x)$ be a matrix of order n whose elements are finite continuous functions of the real variable x on the range $a \leqq x \leqq b$. Let the interval of definition be divided into n segments $h_1, h_2, \ldots, h_n$, and consider the matrices $T_1, T_2, \ldots, T_n$, each T_i corresponding to a value of x contained in the interval h_i. Set $R_i = h_i T_i + I$. Then if the limits

$$\lim_{\text{norm}\, h_i \to 0} R_1 R_2 \ldots R_n, \qquad \lim_{\text{norm}\, h_i \to 0} R_n R_{n-1} \ldots R_1$$

exist, they are called the *right* and *left* RIEMANN *integrals*, respectively, of $T(x)$.

VOLTERRA further showed that if differentiation and integration are both right, or both left,

$$\frac{d}{du}\int_a^u T(x)\, dx = T(u),$$

and that if M is non-singular with elements free from x, then $S(x) = M^{I} T(x)\, M$ implies

$$\frac{d}{dx} S(x) = M^{I}\left[\frac{d}{dx} T(x)\right] M, \qquad \int S(x)\, dx = M^{I}\left[\int T(x)\, dx\right] M.$$

The constant of integration is multiplicative instead of additive. That is, if differentiation is on the right,

$$\frac{d}{dx} C\, S(x) = \frac{d}{dx} S(x), \qquad dC \neq 0$$

where C is a matrix with constant elements; and if, conversely,

$$\frac{d}{dx} T(x) = \frac{d}{dx} S(x),$$

then there exists a non-singular constant matrix C such that $T(x) = C\, S(x)$. If differentiation is on the left, the constant matrix C is a right factor.

VOLTERRA[1] generalized the theorem of CAUCHY to matrices by proving that if s is a closed path inside which each element of $T(z)$ is holomorphic, then

$$\int_s T(z)\, dz = I.$$

He extended this theorem to RIEMANN surfaces.

[1] VOLTERRA: Rend. Circ. mat. Palermo Vol. 2 (1888) pp. 69–75.

L. SCHLESINGER[1] defined the integral

$$\overset{x}{\underset{p}{I}}(a_{rs}) = \int_p^q (a_{rs}\,dx + \delta_{rs})$$

so that every row satisfies the system of differential equations

$$\frac{dy_r}{dx} = \sum_{i=1}^{n} a_{ir}(x)\, y_i .$$

Later[2] he used WEDDERBURN's definition of absolute value (§ 51) to define an L-integrable matrix. He showed that the L-integral of A has a derivative almost everywhere, and that this coincides almost everywhere with A.

H. W. TURNBULL[3] investigated the properties of the operator $\left|\frac{\partial}{\partial x_{sr}}\right|$ applied to matrices whose elements are functions of the x_{rs}.

A. LOEWY[4] defined a new relation between two matrices whose elements are functions of a variable x. Let $P = (p_{rs})$, $P^{\mathrm{D}} = \left(\frac{d p_{rs}}{dx}\right)$. Then we may write $A \overset{\mathrm{Lo}}{=} B$ (A is similar to B in the sense of LOEWY) if there exists a matrix P whose elements are functions of x and whose determinant does not vanish identically such that

$$B = -P^{\mathrm{D}} P^{\mathrm{I}} + P A P^{\mathrm{I}}.$$

It may be shown that this relationship is determinative, reflexive, symmetric and transitive. The author uses it in the factorization theory of differential expressions.

X. Matrices of infinite order.

53. Infinite determinants. The concept of infinite determinant was introduced by G. W. HILL[5] in connection with the solution of differential equations. If

$$A = \left\| \begin{matrix} a_{11} & a_{12} & a_{13} & \cdots \\ a_{21} & a_{22} & a_{23} & \cdots \\ a_{31} & a_{32} & a_{33} & \cdots \\ \cdot & \cdot & \cdot & \cdot \end{matrix} \right\|$$

is a doubly infinite array, and if

$$A_1 = a_{11}, \quad A_2 = \begin{vmatrix} a_{11} & a_{12} \\ a_{21} & a_{22} \end{vmatrix}, \quad A_3 = \begin{vmatrix} a_{11} & a_{12} & a_{13} \\ a_{21} & a_{22} & a_{23} \\ a_{31} & a_{32} & a_{33} \end{vmatrix}, \ldots,$$

then if $\lim_{n\to\infty} A_n$ exists, it is called the determinant $d(A)$.

---

[1] SCHLESINGER, L.: J. reine angew. Math. Vol. 128 (1904) pp. 263—297.

[2] SCHLESINGER, L.: Math. Z. Vol. 33 (1931) pp. 33—61.

[3] TURNBULL, H. W.: Proc. Edinburgh Math. Soc. II Vol. 1 (1927) pp. 111—128.

[4] LOEWY, A.: Math. Ann. Vol. 78 (1918) pp. 1—51, 343—358 and 359—368 — Nachr. Ges. Wiss. Göttingen 1917 pp. 255—263.

[5] HILL, G. W.: Acta math. Vol. 8 (1886) pp. 1—36.

The procedure of HILL was placed upon a rigorous foundation by POINCARÉ[1], who considered only arrays whose diagonal elements are 1's. He proved the existence of HILL's limit under the assumption that $\sum_{p,q=1}^{\infty} |a_{pq}|$ converges.

H. VON KOCH[2] gave an extended and systematic treatment of infinite determinants. If A_n is of order n, set $B_n = (b_{rs}) = A_n - I_n$. Then (Theorem 14.3)

$$|A_n| = 1 + \sum_{i=1}^{n} b_{ii} + \sum_{\substack{i,j=1\\ i<j}}^{n} \begin{vmatrix} b_{ii} & b_{ij} \\ b_{ji} & b_{jj} \end{vmatrix} + \cdots.$$

A determinant of infinite order is said to be *absolutely convergent* if $|A_n|$ and each of the terms in the above expansion converge as $n \to \infty$.

ST. BÓBR[3] proved that a necessary and sufficient condition that $|A|$ be absolutely convergent is that $\prod_{i=1}^{\infty} |a_{ii}|$ converge absolutely, and that there exist an integer $p \geqq 2$ such that

$$\sum_{i=1}^{\infty} \left[\sum_{k=1}^{\infty} |a_{ik}|^p \right]^{\frac{1}{p-1}}, \qquad i \neq k$$

converges. A simplified proof of this theorem was given by L. W. COHEN[4].

Systems of linear equations in infinitely many variables have been discussed by several writers since HILL, for example, E. SCHMIDT[5], TOEPLITZ[6], WINTNER[7] and L. W. COHEN[4].

Application of infinite determinants in the theory of continued fractions was made by T. J. STIELTJES[8] and by VON KOCH[9].

Summaries of results up to the date of publication are given by E. PASCAL[10], KOWALEWSKI[11], F. RIESZ[12], and HELLINGER and TOEPLITZ[13]. The latter has practically complete references to the literature.

[1] POINCARÉ: Bull. Soc. Math. France Vol. 14 (1886) pp. 77–90.

[2] KOCH, H. VON: Acta math. Vol. 15 (1891) pp. 53–63; Vol. 16 (1892–1893) pp. 217–295 — C. R. Acad. Sci., Paris Vol. 116 (1893) pp. 91–93 — Rend. Circ. mat. Palermo Vol. 28 (1909) pp. 255–266.

[3] BÓBR, ST.: Math. Z. Vol. 10 (1921) pp. 1–11.

[4] COHEN, L. W.: Bull. Amer. Math. Soc. Vol. 36 (1930) pp. 563–572.

[5] SCHMIDT, E.: Rend. Circ. mat. Palermo Vol. 25 (1908) pp. 53–77.

[6] TOEPLITZ: Rend. Circ. mat. Palermo Vol. 28 (1909) pp. 88–96.

[7] WINTNER: Math. Z. Vol. 24 (1925) p. 266.

[8] STIELTJES, T. J.: Ann. Fac. Sci. Univ. Toulouse Vol. 8 (1894) J1–J122.

[9] VON KOCH: C. R. Acad. Sci., Paris Vol. 120 (1895) pp. 144–147.

[10] PASCAL, E.: Die Determinanten, trans. by Leitzmann. Teubner 1900.

[11] KOWALEWSKI: Einführung in die Determinantentheorie, Chap. 17. Leipzig 1909.

[12] RIESZ, F.: Les systèmes d'équations linéaires a une infinité d'inconnues. Paris 1913.

[13] HELLINGER and TOEPLITZ: Enzykl. der math. Wiss. II C 13 § 17 (1927).

More recent papers on infinite determinants are by D. C. GILLESPIE[1] and A. A. SHAW[2].

54. Infinite matrices. The modern theory of matrices of infinite order is almost entirely an out-growth of the theory of integral equations. This theory had its inception in a series of six papers by DAVID HILBERT[3] published in the Göttinger Nachrichten under the title "Grundzüge einer allgemeinen Theorie der linearen Integralgleichungen". This was later issued in book form[4]. Subsequent papers by HILBERT, FREDHOLM, HELLINGER, TOEPLITZ, E. SCHMIDT, WINTNER, VON NEUMANN and others have greatly extended the theory.

In spite of the extent and importance of this theory, it will be very briefly treated here because it belongs so fundamentally to the theory of integral equations, and also because it has been so thoroughly expounded in recent years by HELLINGER and TOEPLITZ[5] and by A. WINTNER[6]. Both books have extended references to the literature.

The purpose of our brief remarks on the subject of infinite matrices is to point out a few analogies with the theory of matrices of finite order, and also a few ways in which they are fundamentally different.

55. A matric algebra of infinite order. If $\mathfrak{F}$ is the real or complex field, denote by A_n an array (a_{rs}) of n rows and columns, and by A an array with a denumerable infinity of rows and columns. If multiplication of arrays is defined according to the identity

$$AB = \left(\lim_{n \to \infty} \sum_{i=1}^{n} a_{ri} b_{is}\right),$$

it is at once evident that closure is not usually obtained. Hence it is not possible to define the total matric algebra of order n over $\mathfrak{F}$ for n infinite as was done for n finite.

We shall call a *matric algebra of infinite order* a system of arrays, with elements in a field $\mathfrak{F}$, which satisfies the following postulates[7]:

1. The system is closed under addition.

2. For every two arrays $A = (a_{rs})$, $B = (B_{rs})$, the infinite series

$$\sum_{i=1}^{\infty} a_{ri} b_{is} \tag{54.1}$$

[1] GILLESPIE, D. C.: Bull. Amer. Math. Soc. Vol. 33 (1927) pp. 654—655, abstract only.

[2] SHAW, A. A.: Amer. Math. Monthly Vol. 38 (1931) pp. 188—194.

[3] HILBERT, DAVID: Nachr. Ges. Wiss. Göttingen 1904 pp. 49—91, 213—259; 1905 pp. 307—338; 1906 pp. 157—227, 439—480; 1910 355—417.

[4] HILBERT, DAVID: Leipzig 1912.

[5] HELLINGER and TOEPLITZ: Enzykl. der math. Wiss. II_3 Vol. 9 (1927).

[6] WINTNER, A.: Spektraltheorie der unendlichen Matrizen. Leipzig: Hirzel 1929.

[7] HELLINGER and TOEPLITZ: Nachr. Ges. Wiss. Göttingen 1906 pp. 351—355.

converges absolutely, and the product matrix $AB = (\sum a_{ri} b_{is})$ is in the system.

For some purposes it is convenient to add

3. The system is maximal.

We shall define a *matrix of infinite order* as a member of a matric algebra of infinite order.

There can be more than one maximal system.

Theorem 55.1. Multiplication is associative, and distributive with respect to addition.

Both of these results follow from the assumption that the series (55.1) is absolutely convergent, and hence

$$\sum_{i=1}^{\infty} a_{ri}\left(\sum_{j=1}^{\infty} b_{ij} c_{js}\right) = \sum_{j=1}^{\infty}\left(\sum_{i=1}^{\infty} a_{ri} b_{ij}\right) c_{js},$$

$$\sum_{i=1}^{\infty} a_{ri}(b_{is} + c_{is}) = \sum_{i=1}^{\infty} a_{ri} b_{is} + \sum_{i=1}^{\infty} a_{ri} c_{is}.$$

Theorem 55.2. If A has both a right and a left inverse, they are equal and unique.

Let $$AX = I, \qquad YA = I.$$

Then $$Y = Y(AX) = (YA)X = X.$$

For a matrix of finite order the existence of a left inverse implies the existence of a right inverse and vice versa. That this is not necessarily so for matrices of infinite order may be seen from the following example[1]:

$$A = \begin{Vmatrix} 0 & 1 & 0 & 0 & \dots \\ 0 & 0 & 1 & 0 & \dots \\ 0 & 0 & 0 & 1 & \dots \\ \dots & \dots & \dots & \dots & \dots \end{Vmatrix}, \qquad X = \begin{Vmatrix} x_1 & x_2 & x_3 & x_4 & \dots \\ 1 & 0 & 0 & 0 & \dots \\ 0 & 1 & 0 & 0 & \dots \\ \dots & \dots & \dots & \dots & \dots \end{Vmatrix}.$$

These matrices belong to a matric algebra of infinite order to be defined in the next paragraph. Evidently $AX = I$, while for every Y, YA has only 0's in its first column. It is to be noted that the right inverse of A is not unique, for the x's are arbitrary.

Theorem 55.3. If A has a unique right inverse, this right inverse is also a left inverse[2].

Suppose $AX = I$. Then

$$AXA = A, \qquad AX + AXA - A = I,$$
$$A(X + XA - I) = I.$$

If the right inverse is unique,

$$X + XA - I = X, \qquad XA = I.$$

[1] HELLINGER and TOEPLITZ: Math. Ann. Vol. 69 (1910) pp. 289—330.
[2] HELLINGER and TOEPLITZ: l. c.

A matrix of infinite order is said to be *non-singular* if it has both a left and a right inverse, i.e., if it has a unique inverse.

The failure for infinite matrices of the fundamental theorem of the first chapter (Corollary 7.9), namely that the determinant of the product of two matrices is equal to the product of their determinants, is particularly to be noted. In the example given above with each x equal to zero, we have $AX = I$, while by the definition of determinant due to Hill,

$$d(A) = 0, \quad d(X) = 0, \quad d(I) = 1.$$

By von Koch's definition, $d(X)$ does not exist.

56. Bounded matrices. Let $\mathfrak{F}$ be the real field, and consider all matrices $A = (a_{rs})$ of infinite order such that there exists a positive number m independent of n so that for

$$x_1^2 + x_2^2 + \cdots + x_n^2 \leqq 1, \qquad y_1^2 + y_2^2 + \cdots + y_n^2 \leqq 1,$$

it is true that

$$\left| \sum_{p,q=1}^{n} a_{pq} x_p y_q \right| \leqq m.$$

Such matrices are called *bounded*[1].

Hilbert proved that the product of two bounded matrices exists and is bounded, and that the associative law holds for bounded matrices. Bounded matrices constitute a matric algebra of infinite order[2].

Let $\sum_{i=0}^{\infty} c_i z^i$ be a power series convergent for $|z| < \sigma$. We may consider $1, z, z^2, \ldots$ to constitute a basis for an algebra of infinite order, the constants of multiplication being $c_{ijk} = \delta_{i+j,k}$. (Cf. § 2.) Hence $S_i = (\delta_{r+i,s})$. That is,

$$S_0 = \left\| \begin{matrix} 1 & 0 & 0 & \ldots \\ 0 & 1 & 0 & \ldots \\ 0 & 0 & 1 & \ldots \\ \cdot & \cdot & \cdot & \cdot \end{matrix} \right\|, \quad S_1 = \left\| \begin{matrix} 0 & 1 & 0 & \ldots \\ 0 & 0 & 1 & \ldots \\ 0 & 0 & 0 & \ldots \\ \cdot & \cdot & \cdot & \cdot \end{matrix} \right\|, \quad S_2 = \left\| \begin{matrix} 0 & 0 & 1 & \ldots \\ 0 & 0 & 0 & \ldots \\ 0 & 0 & 0 & \ldots \\ \cdot & \cdot & \cdot & \cdot \end{matrix} \right\|, \ldots$$

The matrix corresponding to $\sum c_i z^i$ is

$$\sum c_i S^i = \left\| \begin{matrix} c_0 & c_1 & c_2 & \ldots \\ 0 & c_0 & c_1 & \ldots \\ 0 & 0 & c_0 & \ldots \\ \cdot & \cdot & \cdot & \cdot \end{matrix} \right\|.$$

As in § 2, these matrices of infinite order are isomorphic under addition and multiplication with the series to which they correspond. Hence the set of all such matrices corresponding to series convergent for

[1] Hilbert: Nachr. Ges. Wiss. Göttingen 1906 pp. 157–227.

[2] Hellinger and Toeplitz: Nachr. Ges. Wiss. Göttingen 1906 pp. 351–355.

$|z| < \sigma$ constitute a matric algebra, and are commutative, associative and distributive.

In fact, if the radius of convergence is $\sigma > 1$, these matrices are bounded[1].

The theory of bounded matrices constitutes the major portion of the known theory of infinite matrices. In particular, all unitary infinite matrices are bounded.

Many of the theorems for bilinear and quadratic forms carry over to forms of infinite order with the understanding that a non-singular matrix is one with a unique inverse[2].

TOEPLITZ[3] gave a method for reducing a bounded quadratic form to canonical form.

H. HAHN[4] gave a necessary and sufficient condition that two bounded quadratic forms in infinitely many variables be equivalent by orthogonal transformations. F. H. MURRAY[5] gave a method for reducing such a form to canonical form by orthogonal transformations.

J. HYSLOP[6] gave an extension of WEIERSTRASS' theorem on pairs of quadratic forms.

I. SCHUR[7] proved that if H is bounded semi-definite hermitian, there exists a P with 0's above the main diagonal such that $H = PP^{CT}$.

A. WINTNER[8] proved that if A is non-singular and bounded, there exists exactly one positive definite matrix P and exactly one unitary matrix U such that $A = PU$.

The theory of infinite orthogonal matrices was developed by M. H. MARTIN[9].

Even if $\sum_{i=1}^{\infty} a_{ri}^2$ converges, A need not be bounded. T. CARLEMAN[10] showed how the study of such matrices may be brought under the theory of bounded matrices.

J. VON NEUMANN[11] considered matrices which are not bounded.

Suppose that Q is a given matrix, and that there exists a matrix P such that $QP - PQ = I$. This condition is never satisfied by matrices of finite order (since the trace of the left member is 0), but is satisfied

[1] TOEPLITZ: Math. Ann. Vol. 70 (1910) pp. 351–376.

[2] HELLINGER and TOEPLITZ: Math. Ann. Vol. 69 (1910) pp. 289–330.

[3] TOEPLITZ: Nachr. Ges. Wiss. Göttingen 1907 pp. 101–109.

[4] HAHN, H.: Mh. Math. Phys. Vol. 23 (1912) pp. 161–224.

[5] MURRAY, F. H.: Ann. of Math. Vol. 29 (1928) pp. 133–139.

[6] HYSLOP, J.: Proc. London Math. Soc. II Vol. 24 (1926) pp. 264–304.

[7] SCHUR, I.: Math. Z. Vol. 1 (1918) pp. 184–207.

[8] WINTNER, A.: Amer. J. Math. Vol. 54 (1932) pp. 145–149.

[9] MARTIN, M. H.: Amer. J. Math. Vol. 54 (1932) pp. 579–631.

[10] CARLEMAN, T.: Sur les équations intégrales singulières à noyau réel et symétrique. Uppsala 1923.

[11] NEUMANN, J. VON: J. reine angew. Math. Vol. 161 (1929) pp. 208–236 — Math. Ann. Vol. 102 (1929) pp. 49—131.

by an important class of infinite matrices used in quantum mechanics. If R is another matrix, then $RP - PR = R'$ is called the derivative dR/dQ of R with respect to Q. From this definition the more important properties of the derivative may be shown to hold for R'.[1]

57. Matrices with a non-denumerable number of rows and colums. Each element a_{rs} of A may be considered as a function of the two variables r and s. From this point of view, every function of two independent variables is a matrix[2].

A matric algebra of non-denumerably infinite order is, then, composed of a set of functions $A = a(x, y), B = b(x, y), \ldots$ which is closed under addition, and under "multiplication", or composition of the type

$$AB = \int_a^b a(x, t)\, b(t, y)\, dt.^3$$

It is readily seen that multiplication is associative and distributive, but usually not commutative[4].

Application of such matrices to physics was made by P. A. M. DIRAC[5].

A. D. MICHAL[6] considered the equivalence of "quadratic forms" of the type

$$\int_a^b\int_a^b g(\alpha, \beta)\, y(\alpha)\, y(\beta)\, d\alpha\, d\beta + \int_a^b [y(\alpha)]^2 d\alpha$$

under FREDHOLM transformations

$$y(i) = \bar{y}(i) + \int_a^b K(\alpha, i)\, \bar{y}(\alpha)\, d\alpha$$

of non-vanishing FREDHOLM determinant. It is here assumed that $y(\alpha)$ and $K(\alpha, \beta)$ are continuous real functions of α and β over the interval (a, b), and that RIEMANN integration is used.

[1] DIRAC, P. A. M.: Proc. Cambridge Phil. Soc. Vol. 23 (1926) pp. 412—418.

[2] MOORE, E. H.: Hermitian Matrices of Positive Type. Lectures at the University of Chicago 1920.

[3] VOLTERRA, V.: Atti Accad. naz. Lincei, Rend. V Vol. 19 (1910) pp. 169—180.

[4] See HELLINGER and TOEPLITZ: Enzykl. der math. Wiss. II C 13 p. 1487.

[5] DIRAC, P. A. M.: Proc. Roy. Soc. London A Vol. 113 (1926—1927) pp. 621 to 641.

[6] MICHAL, A. D.: Amer. J. Math. Vol. 50 (1928) pp. 473—517.

Druck der Spamer A.-G. in Leipzig.